Susan Walker

FLOW CYTOMETRY PRINCIPLES FOR CLINICAL LABORATORY PRACTICE

Quality Assurance for Quantitative Immunophenotyping

FLOW CYTOMETRY PRINCIPLES FOR CLINICAL LABORATORY PRACTICE

Quality Assurance for Quantitative Immunophenotyping

Marilyn A. Owens
Director of Education and Training
Cytometry Associates, Inc.
Dana Point, California

and

Michael R. Loken
Director of Research
Cytometry Associates, Inc.
Seattle, Washington

WILEY-LISS
A JOHN WILEY & SONS, INC., PUBLICATION
New York • Chichester • Brisbane • Toronto • Singapore

Address All Inquiries to the Publisher
Wiley-Liss, Inc., 605 Third Avenue, New York, NY 10158-0012

Printed in the United States of America.

Library of Congress Cataloging-in-Publication Data

Owens, Marilyn A.
Flow cytometry principles for clinical laboratory practice : quality assurance for quantitative immunophenotyping / by Marilyn A. Owens and Michael R. Loken.
p. cm.
Includes bibliographical references and index.
ISBN 0-471-02176-8
1. Immunophenotyping. 2. Flow cytometry—Diagnostic use.
I. Loken, Michael R. II. Title.
[DNLM: 1. Immunophenotyping—methods. 2. Immunophenotyping--standards. 3. Flow Cytometry—methods. 4. Flow Cytometry--standards. QY 250 097f 1994]
QR187.I486094 1994
616.07'582—dc20
DNLM/DLC
for Library of Congress 94-35038
CIP

The text of this book is printed on acid-free paper.

10 9 8 7 6 5 4 3 2 1

Thank you to

David
Beth
Siri
Erik

for their love and support.

Contents

Acknowledgments

This book is an expansion of materials created through Cytometry Associates under contract to the Centers for Disease Control and Prevention, U.S. Department of Health and Human Services. We are grateful to our collaborators in the original project: Gregory T. Stelzer, Andrew Lister, Cindy Morrow, Mark Siragusa, and Linda Stelzer, and to Wallis M. Velleca, CDC Contract Coordinator, who also helped to make that project successful. We thank Cytometry Associates for continuing support and resources in the preparation of this book and the many colleagues and friends who offered discussions and advice on content.

1

Assumptions in Flow Cytometry

INTRODUCTION

Flow cytometry is a powerful technique for obtaining multiparameter data on a wide variety of particles or cells. Its use has been transferred from research settings into the clinical arena, where it has become invaluable as a resource for cell identification, population discrimination, and quantitation.

The clinical utility of this technique is apparent for both qualitative and quantitative applications (Fig. 1.1). The diagnostic procedures that identify malignant populations of hematopoietic cells may be considered *qualitative* evaluations. It is important only to know whether or not the abnormal cell is present. Cell evaluations that require the *quantitation* of populations are becoming more prevalent in routine clinical laboratories. Both types of flow cytometric analysis require that laboratories establish comprehensive quality assurance (QA) programs and maintain documentation of their analyses.

In this book, we will focus on quality assurance for quantitative analyses of cells, where the number, as well as identity, of the cells is important. We will use two clinical examples as illustrative applications: CD4+ T-lymphocyte enumeration in the context of human immunodeficiency virus (HIV) infection and CD34 cell quantitation for peripheral blood stem cell (PBSC) harvests. Other applications will utilize the same principles as discussed here. Although our discussion is limited to blood specimens, cell suspensions from other tissues are amenable to the same types of analysis.

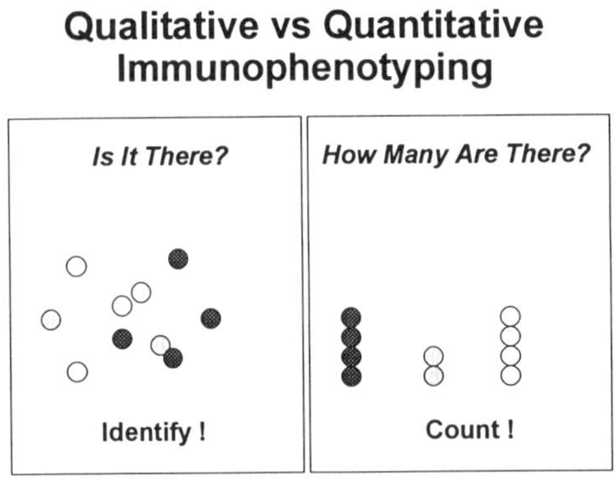

Fig. 1.1. Comparing qualitative with quantitative immunophenotyping. Qualitative applications identify particular subsets of cells, like leukemia cells. Quantitative applications must not only identify the subset of interest, but must also accurately count these cells.

References for each chapter are included in the appendix. Material presented is not to be considered original work by us. It has been assembled to illustrate common themes and generally accepted practice in the clinical setting. These references were selected as a basis for further reading, not as an exhaustive bibliography covering the field. We apologize, in advance, for not including all primary reference materials.

Quality Assurance Assumptions

The quality assurance program for quantitative flow cytometric immunophenotyping (FCI) should be designed to control for all of the sources of error and variability in the FCI procedures. One way to consider these sources of variability is to view the process critically, identifying all assumptions (both stated and implied) that are made prior to the release of patient results.

It is the intention of this chapter to present the facts and stimulate the thought necessary to create an appropriate quality assurance program. The five assumptions inherent in flow cytometric analysis that are described in this chapter will be reconsidered throughout the book.

OVERVIEW OF FLOW CYTOMETRY

The goal of quantitative flow cytometric immunophenotyping in peripheral blood is to enumerate (identify and count) the different types of leukocytes. Blood specimens are collected by venipuncture into tubes containing the appropriate anticoagulant(s). Specimens should be labeled correctly and then transferred to the flow cytometry laboratory for testing. The blood specimen is aliquoted into multiple tubes and stained with different combinations of monoclonal antibodies that are coupled to fluorescent dyes.

The erythrocytes in the stained blood samples are lysed, leaving primarily the leukocytes in suspension. A suspension of cells is then forced, in single file, through the flow cytometer. The cells and the fluorescent dyes that are attached to them interact with laser light when they pass through its beam. The resulting fluorescence emission and scattered light from the cells are collected and converted into electronic signals by the flow cytometer. Flow cytometric data are analyzed using software that assigns cells to various groups and counts the cells with similar characteristics.

Flow cytometers used for FCI identify four or five different characteristics of each cell as it flows through the laser beam. The parameters that are collected for each cell include two or three fluorescence signals, one for each of the dyes used. Light scatter is collected at two angles, including right angle (90°) scatter (SC or SSC) and forward angle (near zero) or forward scatter (FS or FSC). This electronic information is digitized for immediate analysis, or stored in the form of a list (list mode data) of data that can be examined later with a data analysis software program. For example, the number of large, agranular, fluorescent-green cells can be compared with the number of small, agranular, orange-labeled cells.

CLINICAL UTILITY OF QUANTITATIVE IMMUNOPHENOTYPING

Unlike leukemia and lymphoma diagnosis, quantitative immunophenotyping is used primarily for monitoring clinical conditions. Today's needs for quantitative flow cytometry are illustrated in

CD4+ T-lymphocyte enumeration in HIV-infected individuals and in the quantitation of peripheral blood stem cells harvested for transplantation.

CD4 Cell Measurement in HIV Disease

HIV-infected individuals must be monitored at regular intervals to assess the progress of disease. CD4+ T-lymphocyte levels have been determined to be the most appropriate laboratory test for monitoring these patients. Figure 1.2 shows two therapeutic levels of CD4+ cells. When CD4+ T-lymphocyte levels fall below 500 cells/μL, patients may be placed on AZT (zidovudine) or other antiviral therapy. When CD4+ T-lymphocyte levels fall below 200 cells/μL, patients are placed on prophylactic therapy for *Pneumocystis carinii* infection.

Since these recommended intervention points are based on cell quantitation, the quality assurance program for FCI must ensure that patient results are as accurate as possible.

Quantitation of Peripheral Blood Stem Cells

The harvest of stem cells from peripheral blood represents an excellent example of an area of emerging use for quantitative flow immu-

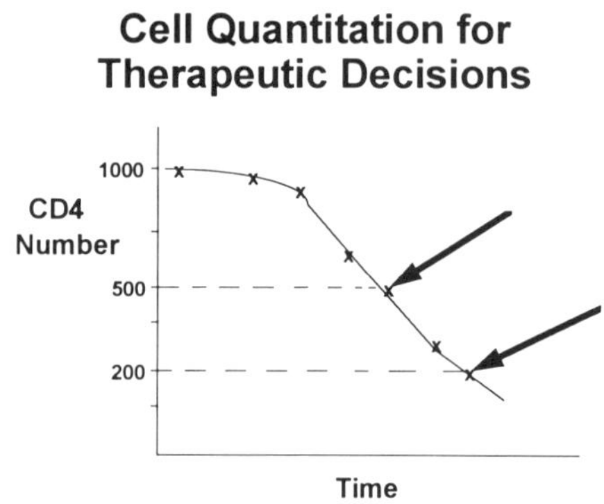

Fig. 1.2. Therapeutic decision points with CD4+ T-cell quantitation. The laboratory measurement of CD4+ T lymphocytes is required for staging individuals with HIV disease. As CD4+ T-cell levels fall to 500 cells/μL (and later, to 200 cells/μL), patients are qualified for antiviral or prophylactic therapies.

nophenotyping. Peripheral blood stem cell harvests are rapidly replacing bone marrow as a source of progenitor cells in the treatment of cancer patients receiving intensified chemotherapy. The cells that will reconstitute the hematopoietic system are destroyed by the high levels of cytotoxic drugs; stem cells are therefore prospectively collected by apheresis and frozen for later transplantation. A successful engraftment requires the transplant of at least 5×10^6 progenitor cells/kg; thus an accurate quantitation of these cells is essential (Fig. 1.3).

An accurate assessment of progenitor cells mobilized by hematopoietic growth factors or chemotherapy is complicated by the extremely low frequency of the cells of interest. Flow cytometric quality assurance is essential to the accurate measurement of these cells and the overall quality control of the cell collection and processing procedures.

QUALITY ASSURANCE AND QUALITY CONTROL

The flow cytometry laboratory's quality assurance (QA) and quality control (QC) programs are established to ensure that correct results

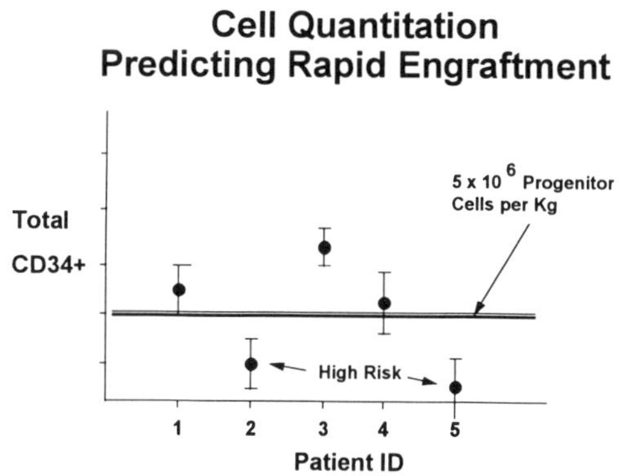

Fig. 1.3. Predicting engraftment success with CD34+ cell quantitation. Peripheral blood stem cell harvests are evaluated for CD34+ cell concentration prior to transplantation. Those patients receiving $>5 \times 10^6$ CD34+ cells/kg have the most rapid rate of engraftment. CD34+ cell enumeration should therefore be part of the QC for PBSC harvests.

are obtained for quantitative cell determinations. The QA program monitors and evaluates the effectiveness of laboratory policies and procedures for the preanalytic, analytic and postanalytic testing phases. The QC program establishes specifications and procedures for (1) performance of the flow cytometer, (2) specimen collection and transport, (3) maintenance of specimen integrity, (4) specimen processing, (5) monoclonal antibody specificity, (6) sample analysis, (7) data acquisition, (8) data analysis, (9) data storage, and (10) reporting correct results.

Understanding the Assumptions

It is important to understand the assumptions made in using flow cytometry. These assumptions concern both cell identification and quantitation, and are the basis for determining whether or not the data are valid. Assumptions are made in the performance of the instrumentation, the specificity of the monoclonal antibodies, the adequacy of sample processing, the ability to logically combine information from multiple sample tubes for a specimen, and the ability to use hematology data together with flow data to obtain absolute counts.

Why is flow cytometry different from other clinical laboratory tests that have evolved from research to routine practice? First, flow cytometry requires a number of steps in a fairly complex procedure. Second, there is less agreement on how to perform the testing than for most clinical laboratory tests. Third, control materials for flow cytometry are not generally quantitative, thus standardization is a relative function. And finally, FCI requires far more operator judgment than routine laboratory tests.

The basic assumptions necessary for FCI are presented below; they will also be discussed throughout the following chapters. The laboratory technologists, supervisor, and/or director should use these assumptions as a guide in developing appropriate QA and QC programs. These assumptions provide the rationale for implementing the various QA and QC procedures.

Validating Underlying Assumptions

Flow cytometric immunophenotyping is a highly complex technology that is comprised of a number of different elements, each presenting a potential source of variation. The inherent biologic variability in the blood specimen is not under the control of the laboratory; however, much of the potential variation in FCI results can be controlled during testing. Identifying and reducing variability in test results is the goal of the QA and QC programs.

There are a number of basic assumptions concerning the variables that are influenced and controlled by the flow laboratory. By stating the assumptions made during immunophenotyping, the basis of a QC program can be defined. Table I.1 lists five areas involving QA that will be discussed throughout this book.

A good QC program tests the validity of each of these assumptions on a daily basis. These assumptions should be used as a reference to test whether or not the procedure or results are sufficient to prove that the assumption is correct. The assumptions should not be considered as correct or incorrect, but simply a necessary part of obtaining accurate FCI results.

Building a Quality Assurance Program

Quality assurance in flow cytometric immunophenotyping implies that all the variables that can be controlled in the analytic process have been monitored and found to be satisfactory. For each of the assumptions made in FCI, quality control assessments must be built into the testing protocol in order to assure accurate results.

ASSUMPTION 1: ADEQUATE SENSITIVITY OF FLOW CYTOMETER

> **It is assumed that the flow cytometer has enough sensitivity to enumerate leukocyte and lymphocyte subsets.**

TABLE I.1. FCI Assumptions

- Instrument sensitivity
- Reagent reproducibility
- Specimen processing consistency
- Sample analytical reproducibility
- Combination of results from different technologies

Light Scatter Resolution

Discrimination between various white blood cells by light scattering must be sufficient to distinguish the lymphocytes and other cell populations in this heterogeneous mixture of cells. Because the populations of cells in blood are so diverse, it is important to remember that the goal is to enumerate subsets of leukocytes and lymphocytes as distinct from cells of other lineages. The ability to distinguish these subsets lies with the optical adjustment of the flow cytometer and the quality of the cell preparation procedure.

Fluorescence Sensitivity

It is assumed that the sensitivity in the fluorescence channels of the flow cytometer allows for the distinction between dim and negative cell populations, that is, the fluorescence channels are sufficiently sensitive to permit the detection of a dimly positive cell as being distinctly different from a negative cell. If the dim cells are counted as negative, they will be underrepresented in the results. If the background is too high or if the instrument is improperly adjusted, negative cells can be counted as positive, thereby overestimating the cell population of interest.

Instrument Stability

It is assumed that the instrument operates identically every time, all of the time. A complex instrument with sophisticated fluidics and five different optical channels may, or may not, operate the same, day in and day out. It is incumbent on the laboratory to ensure that the instrument is operating properly and giving the same results over an extended period of time. Fluctuations in instrument sensitivity or setup can adversely affect the data collected on that day. Quality control procedures must be established to ensure identical operation on a daily basis.

ASSUMPTION 2: CONSISTENT ANTIBODY PERFORMANCE

It is assumed that the monoclonal antibody reactivity is cell-type specific, unambiguous, and reproducible.

Nonspecific Staining

It is assumed that there is minimal nonspecific staining or atypical cell reactivity present in the samples. This, in fact, may be true for normal specimens; however, blood obtained from patients often tends to have increased nonspecific binding when monoclonal antibodies bind to unusual cell types. This type of reactivity usually gives rise to dim, rather than bright cells, so that the distinction between specific versus nonspecific staining can be made based on intensity of fluorescence. This means that the boundaries established to discriminate between positive and negative cell populations must be carefully set.

CD Epitopes

It is assumed that all monoclonal antibodies with the same CD designation are the same. Some antibodies however, react with different epitopes of the CD antigens and therefore may not show the same fluorescence reactivity as other monoclonal antibodies. All reagents used in immunophenotyping must be tested by the laboratory, and any change in clones or vendors will necessitate rechecking the reagents before their use in any clinical testing.

Lot-To-Lot Variability

It is assumed that there is no lot-to-lot variability in monoclonal antibodies. Even though the particular antibody does not change between lots, the ratio of fluorochrome to protein may change in the manufacturing process. It is also possible that there are differences between lots regarding the number of aggregated molecules present. Such manufacturing variability must be tested by the laboratory in order to ensure reproducible staining when production lots of antibodies are changed.

ASSUMPTION 3: MINIMAL VARIABILITY INTRODUCED DURING SAMPLE HANDLING AND PROCESSING

It is assumed that variations introduced during sample processing and analysis do not affect the results.

Specimen Collection, Transport, and Storage

It is assumed that the data are not biased by specimen collection, shipping, or storage. Once the cells leave the body and are collected in the vacuum tube, the specimen begins to deteriorate. The rate of deterioration depends on the anticoagulant used to collect the specimen, the temperature at which the specimen is stored, and the length of time the specimen sits before it is processed. For this reason, it is essential to standardize the shipping and storing procedures and to select the proper anticoagulant for specimen collection. It is preferable for samples to be analyzed within 30 hours after collection. Any specimen over 48 hours old must be rejected.

Specimen Processing

It is assumed that specimen processing does not introduce variability in the quantitative results. Different methods of processing bias the results in different ways. It is important to know how the specimen processing affects the data. This also means that it is important to be consistent in the method used to process the specimen so that the results can be compared with other results obtained at a different time.

Data Analysis

It is assumed that data analysis does not affect the subset proportions. The setting of markers and gates to analyze heterogeneous populations also introduces bias into the data. Changing the light scatter analysis gate that identifies lymphocytes can include or exclude certain subsets of lymphocytes, as well as other types of cells. Small changes in fluorescence marker settings can also change the assignment of some cells as positive or negative. This means that the methods used for analyzing the data must be consistent from sample to sample and specimen to specimen. This consistency is essential to minimize the variability of results.

ASSUMPTION 4: LOGICAL COMBINATION OF RESULTS

It is assumed that subset information from different samples of the same specimen makes logical sense and that data from different reagent combinations can be logically combined.

Logical Relationships

Since flow cytometric immunophenotyping is often based on results from several reagent combinations, the results of different sample tubes must fit a defined pattern. Reagent combinations may be used to define logical relationships between subsets; for example, the sum of all lymphocyte subsets should equal the total number of lymphocytes. The isotype control should have fewer positive cells than the positive control. Throughout the analysis of a particular specimen, the logical relationships between subsets must be maintained.

Common Light Scatter Patterns

It is assumed that common characteristics between samples are maintained. For example, if a specimen has been aliquoted and stained with a variety of antibodies, the common feature for all samples is the cellular composition and its associated light-scattering characteristics. Since the samples are processed in separate tubes, however, errors may occur in the processing so that the light scatter patterns and detected cellular composition may change from one sample to another. It is important to monitor these changes, since they are indicators of problems with either the specimen or the sample tube(s).

Light Scatter Distinguishes the Lymphocytes

For CD4+ cell enumeration, it is assumed that lymphocytes can be uniquely and completely identified by light scatter. The FCI analysis is performed to distinguish between the lymphoid subsets; however, because of the heterogeneity of cells in a blood sample, it may be difficult to distinguish the lymphocytes from the other cell types. Therefore, it may not be possible to isolate lymphocytes completely from other cells present in the specimen. It is also important to remember that not all lymphocytes scatter light identically.

The discrimination of lymphocytes using light scatter is performed with two opposite and often conflicting goals. First, it is necessary to include all of the lymphocytes to be analyzed within an analysis window or gate. On the other hand, it is also important to have the window comprise only lymphocytes and therefore reflect high purity. Setting a larger gate includes all lymphocytes but also includes nonlymphoid cells in the analysis, while setting a smaller analysis window increases the purity of the cells being analyzed but may in fact exclude certain specific subsets of cells. For this reason, it is important to carefully define the procedures for identifying lymphocytes based on light scatter and setting the light scatter gates.

For CD34 analysis, parameters other than light scattering must be used to focus on the appropriate cells. The combination of characteristics must allow detection of all progenitor cells present in the specimen.

Minimal Variation Between Sample Tubes

It is assumed that the variation between samples is minimal, random, and does not affect the data. Since different samples from the same specimen are processed separately, there is always a possibility that one sample may be handled differently than the others. It is important to assess the potential variability produced by separate sample processing, both in the assessment of the light scattering pattern, and in the determination of the intensity and frequency of antibody staining. Replicate staining provides for a measure of variability from sample to sample.

ASSUMPTION 5: LYMPHOCYTE PERCENTAGES FROM FLOW CYTOMETRY AND HEMATOLOGY ARE EQUIVALENT

It is assumed that the proportion of leukocytes or lymphocytes identified by either a flow cytometer or a hematology analyzer is identical.

Hematology Analyzer Reproducibility

It is assumed that there is no variation between different commercial hematology instrumentation systems in determining the absolute counts of lymphocytes under all conditions. It is well documented that different hematology analyzers identify leukocytes in different ways. The optical or electronic principles used in the different instruments result in the cells being viewed differently. The different procedures to lyse the erythroid cells also introduce a bias into the data. This means that there can be significant variation introduced in the immunotyping analysis simply due to the white blood cell (WBC) and/or differential counts.

Comparison of data from one site to another may require standardization of the hematology instruments, especially when absolute counts are being compared. This may be true even for different hematology instruments in the same facility, and such differences may account for the majority of variations in results for identical specimens at the same facility.

Flow Cytometer Reproducibility

It is assumed that all flow cytometers identify cell populations identically. In fact, instruments from different manufacturers, as well as different models from the same manufacturer, have slightly different optical configurations and differing abilities to distinguish cells based on light-scattering characteristics. In addition, there are significant differences in sensitivity between different instruments. Therefore, data obtained using one instrument may be slightly different than data obtained on another instrument with the same specimen.

Different Brands of Flow Cytometers

It is assumed that the lymphocyte proportions determined by two different brands of flow cytometers should always be identical. Comparison of data from laboratory to laboratory, however, may be affected by differences in the optical characteristics of the different flow cytometers. Care must be taken in assuming that the same reference range and test results can be obtained when using more than one instrument.

QUALITY ASSURANCE

An integrated quality assurance program must be designed to validate all of the assumptions inherent in obtaining flow cytometric results. The awareness of which variables can be monitored is important in understanding how the QA program should be structured. Once the program is established, it is essential to follow quality control criteria in order to minimize bias in the results and to provide consistent data that can be compared over time.

CONCLUSIONS

The goal of quantitative immunophenotyping by flow cytometry is to identify and to enumerate subsets of leukocytes in peripheral blood. In order to develop an effective quality assurance program for this testing, laboratory operators must consider the assumptions made in flow cytometric immunophenotyping and build quality control specifications to match. Assumptions regarding instrumentation, reagents, sample processing, analysis, and calculations will be discussed throughout this book.

FURTHER READINGS

Centers for Disease Control. CDC 1993 revised classification system for HIV infection and expanded surveillance case definition for AIDS among adolescents and adults. *MMWR Morb Mortal Wkly Rep* 41(RR-17):1–19.

Givan AL. *Flow Cytometry: First Principles*. Wiley-Liss, New York, 1992.

Lister A, Siragusa M, Loken MR, Stelzer GT, Owens MA, Morrow C, Stelzer L. *Flow Cytometric Immunophenotyping Quality Assurance and Quality Control Procedures for T Lymphocyte Subset Enumeration: Procedure Manual*. U.S. Department of Health and Human Services, Centers for Disease Control and Prevention, Atlanta, 1993.

Loken MR, Civin CI. Laboratory assessment of lymphohematopoietic progenitor and stem cells. In Smith DM, Sacher RA, eds. *Peripheral Blood Stem Cells*. American Association of Blood Banks, Bethesda, MD, 1993, 19–31.

Loken MR, Stall AM. Flow cytometry as an analytical and preparative tool in immunology. *J Immunol Methods* 1982, 50:85–112.

Loken MR, Stelzer GT, Owens MA, Morrow C, Lister A, Siragusa M, Stelzer L. *Flow Cytometric Immunophenotyping Quality Assurance and Quality*

Control Procedures for T Lymphocyte Subset Enumeration: Self Study Guide. U.S. Department of Health and Human Services, Centers for Disease Control and Prevention, Atlanta, Course 3019-G, 1993.

McCarthy RC, Fetterhoff TJ. Issues for quality assurance in clinical flow cytometry. *Arch Pathol Lab Med* 1989, 113:658–666.

Willman C, Stewart C. General principles of multiparameter flow cytometeric analysis: applications of flow cytometry in the diagnostic pathology laboratory. *Semin Diagn Pathol* 1989, 6:3–12.

STUDY QUESTIONS

1. Distinguish between *qualitative* and *quantitative* immunophenotyping.
2. Name one example of qualitative immunophenotyping.
3. List two examples of quantitative immunophenotyping for clinical use.
4. What is the "goal" of quantitative flow cytometric immunophenotyping?
5. List at least five specifications and procedures which should be part of a FCI laboratory's QA and QC programs.
6. List four ways that flow cytometry is different from other clinical laboratory tests.
7. Describe assumptions about instrument sensitivity that must be checked in a quality assurance program.
8. Explain how the sensitivity of a flow cytometer is tested.
9. Why is instrument stability so important?
10. What are the two components of "nonspecific staining?"
11. Describe assumptions about monoclonal antibodies that must be monitored in a quality assurance program.
12. Describe assumptions about specimen quality and processing that must be controlled in a quality assurance program.
13. Describe assumptions about combining results from different sample tubes that must be evaluated in a quality assurance program.
14. Explain the difficulties in setting a light scatter gate for lymphocytes.

15. Describe assumptions about comparable data from hematology and flow cytometry that must be validated in a quality assurance program.
16. Why is it important for FCI laboratories to follow the same procedures?

II

Components of Flow Cytometric Immunophenotyping

INTRODUCTION

Flow cytometry involves the identification of cells by a combination of their light-scattering properties and surface antigen binding to fluorochrome-conjugated monoclonal antibodies. All components of flow cytometric immunophenotyping must be monitored to assure the accurate identification and enumeration of cells by these parameters. Each of the components—optics, fluidics, sample preparation, electronics, and data processing—must be evaluated in the development of a quality assurance program.

FLOW CYTOMETRY AND MICROSCOPY

Flow cytometry and microscopy are both techniques that can be used to quantitate cells with certain identifying markers. There are many similarities between these techniques, but also several specific differences.

Similarities

Both microscopy and flow cytometry require that all of the optical components be properly aligned to optimize the detection and classification of each cell. Optical filters are used, both in the fluorescence microscope and in the flow cytometer, to distinguish color differences between fluorochromes or fluorochrome-stained cells. All optical components must be properly focused to maximize the light falling on the cells and the resolution at the ocular. In each case, it is necessary to place the cell of interest in the center of the illuminating light beam and the light-collecting optics.

Differences

Although similar in their components, flow cytometry and microscopy are different in the type of light used, the way that morphology is assessed, the quantitation of fluorescence, and the number of cells counted.

The flow cytometer uses monochromatic laser light rather than incandescent light to illuminate the cell. This laser light has a single wavelength, and is highly focused, and polarized. The fluorescence microscope lamp generally transmits light of many wavelengths that have to be separated by optical filters and focused prior to excitation of the fluorochrome-labeled cells. Flow cytometric cell morphology is determined by observing the light scatter differences between cells.

The flow cytometer obtains readings for fluorescence intensity and can provide quantitative measurements of cell staining. The fluorescence microscopist can make quantitative observations about morphology and fluorescence staining; however, these are subjective rather than objective. The microscopist has an advantage in being able to observe the physical relationships of the cells with one another. Flow cytometers cannot determine the location of the fluorescence on the cell and cannot distinguish between diffuse, patchy, or capped staining. Spatial relationships between cells that are important to pathologists or morphologists are lost in flow cytometry.

Large numbers of cells can be analyzed by flow cytometry, up to 1000 cells-sec. While the flow cytometer frequently analyzes more than 10,000 cells per sample tube in a short amount of time, the microscopist usually observes only several hundred cells per slide over a comparatively longer time span. This means that infrequent cell populations can be enumerated by the flow cytometer with greater efficiency and much higher statistical confidence than can be done by manual counting using a microscope.

BASICS OF FLOW CYTOMETRIC IMMUNOPHENOTYPING

Flow cytometric immunophenotyping follows several analytical steps: Cells are first identified by light-scattering properties, then further defined by the binding of fluorochrome-tagged monoclonal antibodies. The power of this technique comes from multiparameter analysis of each cell. By combining all light scatter and fluorescence information about each and every cell, populations of cells with common characteristics can be identified and quantified.

Displaying Flow Cytometric Data

Data can be displayed as a single-parameter histogram or correlated two-parameter dot plot display, as shown in Figure 2.1. Data may also be shown in contour plots and three-dimensional displays.

Single-parameter histogram. A single-parameter histogram is a probability curve in which cell number or number of events is plotted relative to the intensity of the signal (Fig. 2.1, top). The histogram is a frequency distribution of how many of the events occur at a particular intensity.

Correlated two-parameter display. Correlated two-parameter displays (often termed two-dimensional histograms) depict the relationship between two independent cellular characteristics.

1. Dot plots represent each event as an individual dot (Fig. 2.1, bottom). Groups or clusters of dots identify particles with similar characteristics. With this type of display *infrequent clusters* or events become more obvious and are therefore emphasized relative to major populations.
2. Contour plots provide frequency information for two parameters. Contour plots are similar to contour maps, which define topography or height differences.

Multidimensional display. It must be kept in mind that four or five separate characteristics are detected simultaneously on the flow cytometer. Visualization of such complex data is difficult. Often, several different projections of parameters taken two at a time are used to identify cell groups (Fig. 2.2).

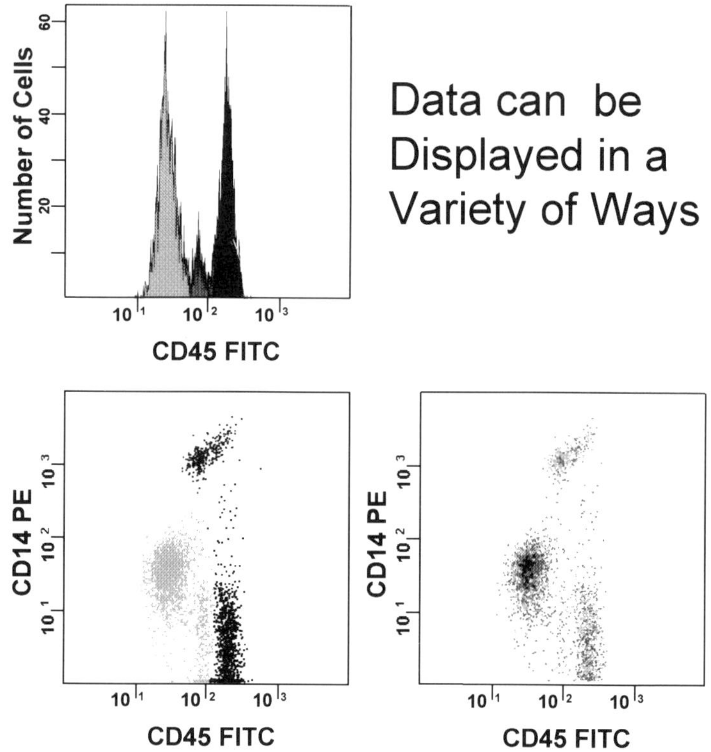

Fig. 2.1. Displays of flow cytometric immunophenotyping data. Two of the most common types of data display in flow cytometry are the single-parameter frequency histogram and the dual-parameter dot plot. The addition of the second parameter allows resolution of populations not visualized in single-parameter displays. Data is expressed either as number of cells or channel number of fluorescent signal.

Multidimensional, color-related displays can then be used to observe relationships between more than two signals at a time. In these displays, populations of cells are given a color (or highlight) that is maintained in other projections of the data.

Morphologic Analysis by Light Scattering

Cell lineages can be identified by their light-scattering properties when analyzed by flow cytometry. Light scattering can be used to differentiate cells by size, shape, or internal complexity (granularity). A combination of forward angle light scatter (FSC or FS) and

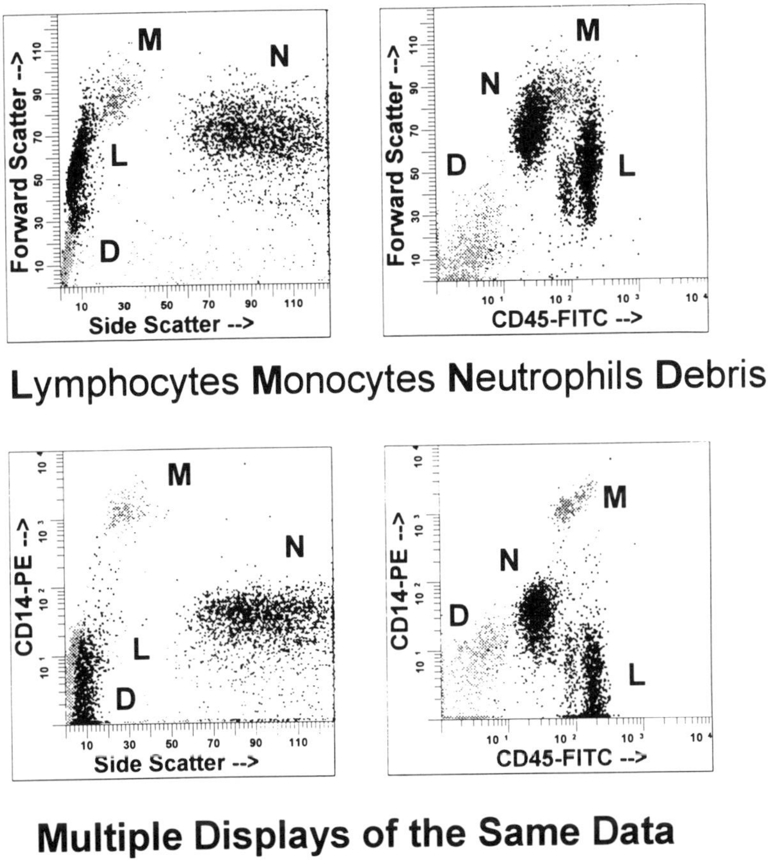

Fig. 2.2. Displays of multidimensional data. When more than two parameters of data on a single sample must be evaluated, histograms are reviewed for all parameters in pairs. Thus, forward light scatter and side scatter may be displayed together in a two-parameter plot and also displayed in plots with fluorescence parameters. In each display, cell populations can be identified as clusters and assigned a color to be related to other clusters in other two-parameter displays.

right angle light scatter (90° SC, side scatter, SSC, or SC) is collected and analyzed for each cell (Fig. 2.3).

Forward light scatter has been shown to be related to cell size, while right angle light scatter corresponds to cell shape or internal complexity. Lymphocytes have low forward- and right-angle light scatter because they are small, smooth, and have little granularity.

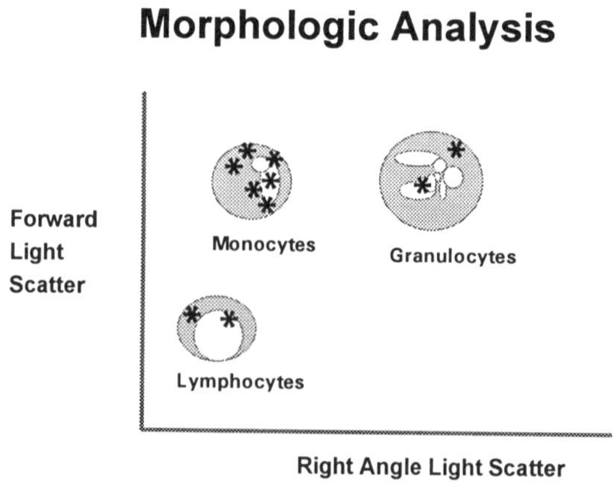

Fig. 2.3 Light scatter of lymphocytes, monocytes and granulocytes. Fluorescent signals (*) are not detected.

Granulocytes are large and have multiple nuclei with large numbers of granules and, therefore, scatter more light in both forward and right angle directions. Monocytes are larger than lymphocytes so they scatter more light than lymphocytes; however, they are less granular than the granulocytes and therefore have intermediate right angle light scatter.

Light scatter display. The light scatter display can either be made with right angle light scatter plotted along the x-axis and forward light scatter along the y-axis or with forward light scatter on the x-axis and right angle light scatter on the y-axis (Fig. 2.4). Both displays give the same pattern discrimination for identifying cell populations.

Antigen Detection by Immunofluorescence

Protein, carbohydrate, or lipid determinants on the surface of cells can be used to identify specific cell types. Monoclonal antibodies, defined by their binding specificity for these determinants, can be tagged with fluorescent dyes. Excitation of the fluorescent dyes by a monochromatic light source will elicit a characteristic spectrum of emitted light, visible by using either a fluorescence microscope or a flow cytometer. Each cell is then assessed for the presence or absence of fluorescent signal emissions, which are related to the cell surface membrane characteristics of that cell.

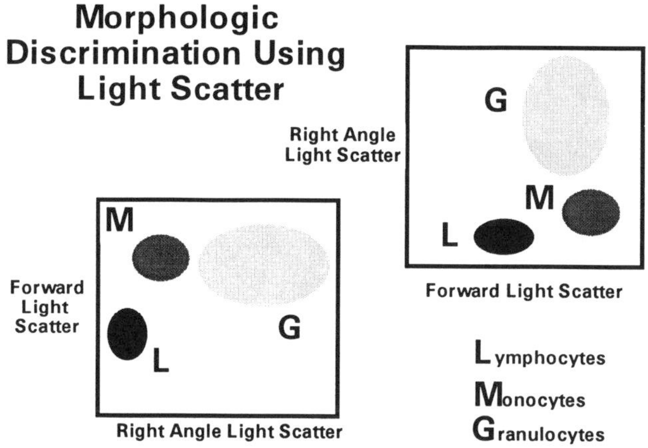

Fig. 2.4. Two orientations for light scatter displays. Regardless of whether forward light scatter or side scatter is displayed on the y-axis of two parameter light scatter plots, the same population patterns can be discriminated and cells can be identified as lymphocytes, monocytes, or granulocytes.

Multiparameter Cell Analysis

The flow cytometer is able to evaluate each cell for several parameters simultaneously, making this technology perfectly suited for multiparameter analyses. Forward angle and right angle light scatter, along with the specific wavelengths and intensities of fluorescence, are measured by the flow cytometer's photodetectors.

Depending on the flow cytometer configuration, four or more different cell characteristics can be evaluated at the same time when monocolonal antibodies are conjugated to different fluorochromes. The flow cytometer must be calibrated to separate and to quantify specific fluorescence emission from each dye, and to do the same for scattered light.

The discussion in this book will be limited to single-laser excitation with two to three fluorescence detectors. The usual fluorescent dyes for two-color immunofluorescence are fluorescein isothiocyanate (FITC) and phycoerythrin (PE). The FITC fluorochrome emits green light and PE emits orange light when excited by the flow cytometer's laser beam. A variety of third color fluorochromes, emitting red light, may be utilized. These include tandem conjugates of multiple dyes, combining properties of each, or Peridinin-chlorophyll *a* (PerCP). The vital stain 7-amino-actinomycin D(7-AAD) may also be used as a third chromophore to distinguish between live and dead cells.

IDENTIFYING LYMPHOCYTES FOR SUBSETTING

Although they are morphologically indistinguishable, lymphocyte subsets are functionally heterogeneous. The major types of lymphocytes include B lymphocytes, T lymphocytes (which are further subdivided into helper/inducer and suppressor/cytotoxic cells), and natural killer (NK) cells.

Cells belonging to different lymphocyte subsets may scatter light slightly differently, as shown in Figure 2.5. Therefore, the composition of cells within the lymphocyte light scatter region is not homogeneous. It should also be noted that activated lymphocytes appear larger than resting lymphocytes because they exhibit greater forward light scattering. They are as large as or larger than the NK lymphocytes.

Nonlymphocyte Light Scatter

Other cells or particles in a prepared flow cytometric immunophenotype (FCI) specimen can scatter light like a lymphocyte. Such

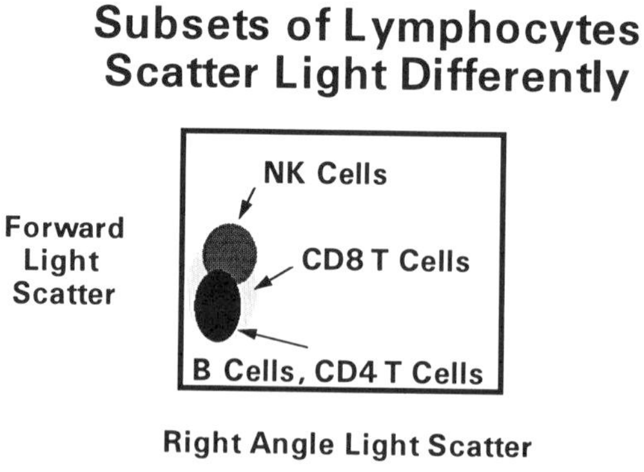

Fig. 2.5. Heterogeneity of lymphocytes. Lymphocytes in different subsets will scatter light differently. For example, NK cells are large and generally scatter the most light of any lymphocytes. B cells and some CD4+ T cells are small and have the least light scatter. CD8+ T cells are heterogeneous in their light scatter.

nonlymphocyte contamination can present problems in lymphocyte subset analysis. Based on their light-scattering properties, basophils are almost indistinguishable from lymphocytes, and erythrocytes and aggregated platelets, and dead or dying neutrophils or monocytes may also be mistaken for lymphocytes.

Identifying Lymphocytes by Immunofluorescence

It may be impossible to identify lymphocytes solely by their light-scattering signal. Distinguishing lymphocytes from other cell types is dependent upon many factors and immunofluorescence may be required to accurately define lymphocytes.

Identifying Progenitor Cells by Immunofluorescence

Progenitor cells represent 0.1–5.0% of peripheral blood in patients following stem cell mobilization; thus their quantitation must be considered as rare event analysis. Obviously, light scattering alone is not adequate to enumerate progenitor cells. In addition to using immunofluorescence for positive identification of progenitors, fluorescence must be used to exclude dead cells and other more mature myeloid cells. Nonspecific staining must be carefully titered for this analysis, as a very small overestimation can dramatically influence the results.

Resolution of Cell Populations

The focusing of the laser beam and the size of the beam stop dramatically affect the forward light scatter signal. Therefore, significant differences can be observed both between instruments, and from day to day on individual instruments in which these optical components can be adjusted. Such potential for variability must be assessed during QC checks of the system.

Resolution of cell populations is dependent on proper optical alignment of the flow cytometer (Fig. 2.6). A dramatic change in light scattering resolution on a single instrument suggests a change in the optical characteristics of that instrument. This may require realignment of the instrument or a service call by a trained representative.

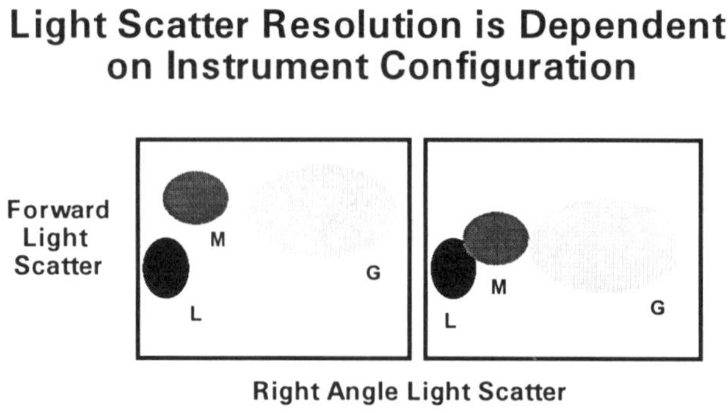

Fig. 2.6. Light scatter resolution for QC of optical alignment. Each flow cytometer must be characterized for its expected light scatter resolution pattern. If any changes in the light scatter pattern occur, the optical components should be realigned.

Population Discrimination

Combining all flow cytometric information for each cell, including light scattering and fluorescence, provides the data that can be analyzed by the computer's hardware and software. Populations of cells can be distinguished by their characteristic light-scattering patterns. Further discrimination of cell populations is possible based on the presence and quantity of fluorescent dye that is bound to the cell surface via the monocolonal antibody reagents. For example, the CD45, CD14 reagent combination is used to distinguish lymphocytes from other leukocytes (Fig. 2.7): All leukocytes are stained with CD45 FITC monoclonal antibody reagent, but at different levels of intensity. Lymphocytes stain brightly with CD45 antibody reagent, but scatter little light due to their small size. Monocytes scatter more light, but are not stained as brightly with CD45. Granulocytes scatter the most light, but are less brightly stained with CD45. As a further distinction, monocytes stain brightly with the CD14 PE monoclonal antibody reagent, while lymphocytes are unstained and some granulocytes are weakly stained.

Populations of cells that cannot be resolved by one or two parameters may be completely separated by the addition of a third, fourth, or fifth characteristic. Visualization of such complex data becomes difficult.

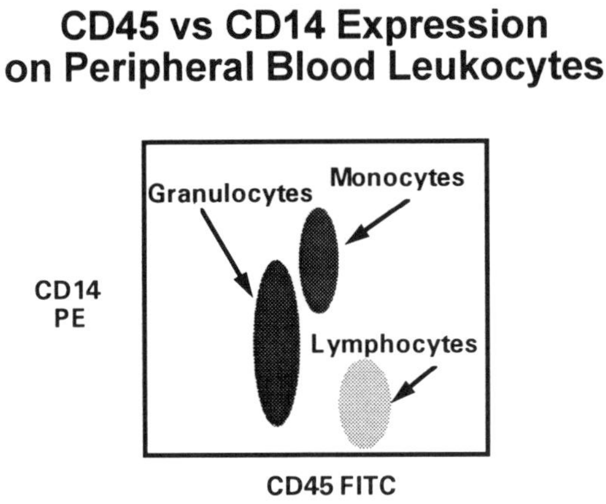

Fig. 2.7. Fluorescence patterns for peripheral blood leukocytes stained with CD45 FITC and CD14 PE. Three populations of peripheral blood leukocytes are easily distinguished with CD45 and CD14 immunophenotyping. Subpopulations can be identified with additional monoclonal antibodies and multiple two-parameter data displays.

By defining specific light scatter and fluorescent electronic gates, many subsets of each cell population can be identified with specific monoclonal reagents. These subsets may be particularly important in diagnostic flow cytometry.

IMMUNOFLUORESCENCE

In immunofluorescence the specificity of binding is conferred by the binding site (B) of the antibody shown in Figure 2.8. The fluorescent dye is simply a tag on the antibody and does not bind to the cell in a specific manner.

This binding is different from that of DNA analogs such as propidium iodide, or other cellular probes in which the specificity and fluorescent properties of the dye must be combined into the same molecule. The separation of binding specificity and spectral properties means that different dyes can be conjugated to identical antibodies and that the same dye can be conjugated to different antibodies.

When using the same dye combinations, the same optical filters and optical configuration can be used on the flow cytometer, even

Immunofluorescence Separates Specificity and Dye Properties

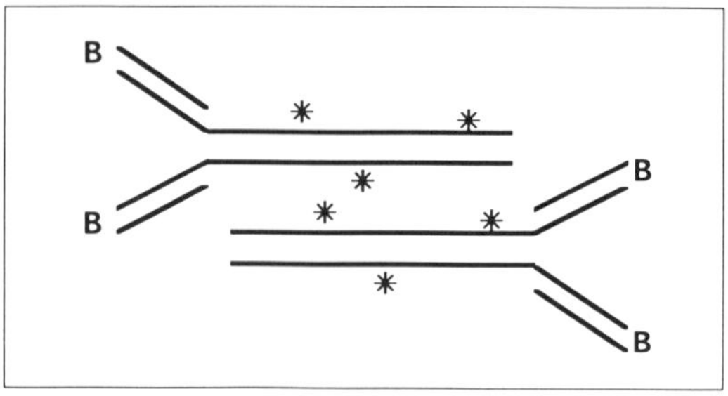

Fig. 2.8. Fluorescent dyes conjugated to immunoglobulins. The specificity of reactivity is conferred by the antibody binding sites (B). Since the fluorescent markers do not affect antibody binding, different monoclonal antibodies can be marked with the same fluorochrome or one monoclonal antibody may be conjugated with any of several common fluorescent dyes.

when the antibody combinations are changed from one sample to another. Different cellular characteristics can be probed simply by changing the antibodies or combination of antibodies used in the sample.

Nonspecific Staining

Nonspecific binding can become a problem when the antibodies attach to cells via mechanisms that do not involve the specific binding site. Such nonspecific binding can occur with antibody aggregates formed during manufacture or storage of the reagent. The Fc receptors of cells (especially monocytes) bind different classes of immunoglobulins and may provide a nonspecific component to specific staining by the monoclonal antibodies.

Surface charges on cells can result in nonspecific staining since conjugation of fluorochromes to antibodies changes the net charge of the monoclonal antibody. This nonspecific binding is particularly a problem with nonviable or dying cells.

Fluorescent Dyes

Characteristics of some of the most commonly used fluorescent dyes in flow cytometry are listed in Table II.1.

Fluorescein. Fluorescein isothiocyanate (FITC) is a small charged molecule, which can be easily conjugated to proteins through an

TABLE II.1. Characteristics of Fluorescent Dyes

	FITC	R-PE	PerCP	Tandem*	7AAD
Ex Max	495 nm	550 nm	490 nm	550 nm	~500 nm
Em Max	530 nm	575 nm	677 nm	670 nm	660 nm
ε 488	86,000	2,000,000	300,000	2,000,000*	?
Quantum Yield	0.5	1.0	1.0	??*	0.035
MW	365	240,000	35,000	240,000	1270

* (R-PE:Cy5); effective extinction coefficients and quantum yields are difficult to establish.

Ex Max = Excitation maximum wavelength; Em Max = Emission maximum wavelength; ε 488 = Extinction coefficient at 488 nm; MW = molecular weight.

isothiocyanate group. It has a high extinction coefficient and its absorption maximum is matched to the output of an argon ion laser (488 nm). Fluorescein has a high quantum yield and approximately half of the absorbed photons are emitted as fluorescent light.

The emission maximum of fluorescein is at 530 nm so that the fluorescence is green to the eye. However, significant amounts of yellow and orange light also come from the fluorescein molecule. Fluorescein is very stable if neutral pH is maintained, but fluorescence is reduced under acidic conditions.

Phycoerythrin. Phycoerythrin (PE) is a large molecule (bigger than IgG) that occurs naturally as part of the photosynthetic apparatus of bacteria, algae, and plants.

Phycoerythrin has an exceedingly high extinction coefficient at 488 nm, even though its peak absorption occurs at 540 nm. Two forms of phycoerythrin (B and R) are obtained from different biologic sources (bacteria and blue-green algae) and have slightly different absorption spectra. This molecule has an extremely high quantum yield, approaching one photon emitted for every photon absorbed.

The conjugation of phycoerythrin to antibodies is much more complex than that of fluorescein. The fluorescent properties are relatively independent of pH, although at extreme pH levels the molecule will dissociate.

Peridinin-chlorophyll *a* complex. Peridinin-chlorophyll *a* complex (PerCP) is a biologic pigment obtained from the photosynthetic systems of dinoflagellates. This dye is a water-soluble protein with a molecular weight of 35,000. The chromophore can be excited at 488 nm, with emission peaking at 680 nm. Monoclonal antibodies can be conjugated to PerCP to provide a third color for immunofluorescence when used with FITC and PE.

The major advantage of using this dye over tandem conjugates is the constancy of the emission spectrum and the limited spectral overlap with emission from PE. As a result, compensation adjustment between PE and PerCP is minimized and constant from lot to lot of reagent. Tandem conjugates have a larger compensation correction, which is variable and depends upon the conjugate efficiency.

Antibodies coupled to PerCP work well with flow cytometers utilizing small argon ion lasers. When stimulated with more than 30–50 mW of 488-nm laser power, the dye can photo-bleach. This results in little or no apparent staining when analyzed on a large laser system although bright fluorescence is observed using a small laser. The PerCP dye is light-sensitive and must be kept in the dark to reduce photodecomposition.

Tandem conjugates. These dyes are conjugates of two dyes that combine their spectral properties. Originally, PE-allophycocyanin and PE-Texas red conjugates were used. Problems in production and reproducibility forming stable dyes with these combinations have given rise to the use of PE-Cyanin 5 couplet. The donor chromophore in tandem conjugates absorbs light and transfers the energy to the other dye, which subsequently emits the light. The combination has the spectral properties of both dyes, with excitation and emission spectra well separated.

The tandem conjugates, by their very nature, have significant spectral overlap with PE and require large compensation settings. The amount of spectral overlap varies from lot to lot and depends upon the efficiency of conjugate formation and subsequent energy transfer from donor to acceptor. This necessitates the setting of the compensation network for each dye lot used in the analysis. Three different antibodies conjugated with tandem dyes may have three different compensation settings, one for each antibody. This increases the need for careful QC of the instrument setup.

Tandem conjugates of PE-Cyanin 5 also have increased nonspecific binding to monocytes. This property must be taken into account during the analysis of the data.

7-Amino-actinomycin D. Dead cells can be distinguished from live cells using 7-amino-actinomycin D (7-AAD) in a manner similar to the discrimination using propidium iodide. The 7-AAD increases its fluorescence upon binding to double-stranded DNA. This is a relatively large molecule, which is excluded from cells with intact plasma membranes. When a cell dies, the outer membrane is ruptured, allowing free passage of this molecule, which can then bind to the DNA and fluoresce brightly.

The emission of 7-AAD is at 660 nm, so there is little spectral overlap with PE. This permits two-color immunofluorescence analysis on cells that are stained to distinguish live and dead cells. The spectral properties of propidium iodide with emission at 620 nm limits its use as a vital stain because of the spectral overlap with PE.

Emission Spectra

Figure 2.9 shows the emission specta of FITC, PE, and PerCP. These wavelength peaks are separate and are used for choosing optical filters for the flow cytometer. All three chromophores can be excited at the same wavelength, 488 nm. These dyes have different emission spectra so they can be distinguished based on their color.

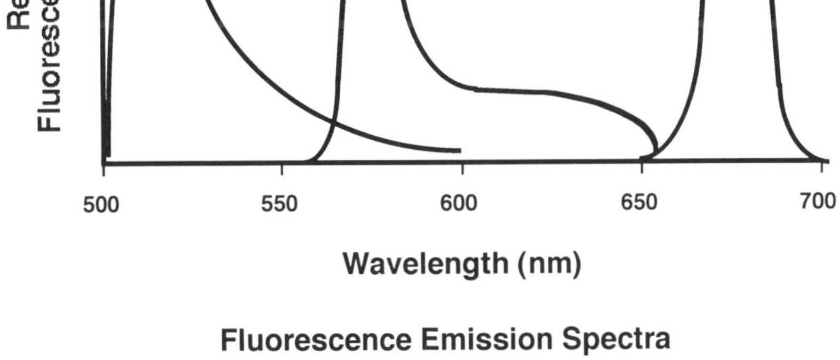

Fig. 2.9. Emission spectra for three common fluorochromes. FITC, PE, and PerCP can all be exited at 488 nm, but emit light in different regions. Any overlap in these emissions should be corrected by spectral compensation.

Optical filters must be selected to block any laser light (488 nm) scattered into the fluorescence detectors but to transmit fluorescent light coming from the cell. The optical filters must discriminate between the emission from each of the chromophores. Optical filters are not perfect, so some light from fluorescein enters the phycoerythrin channel and vice versa. More fluorescein emission enters the phycoerythrin channel than phycoerythrin emission enters the fluorescein channel.

Spectral cross talk can be corrected electronically by a *spectral compensation network*. This system electronically subtracts a portion of the unwanted green (FITC) signal from the PE (orange) detector and the unwanted orange signal from the FITC (green) detector. Similarly, a compensation network corrects for spectral overlap between the orange and red fluorescence channels.

Sensitivity and Autofluorescence

The sensitivity of the flow cytometer is usually limited by the autofluorescence signal from unstained cells as shown in Figure 2.10.

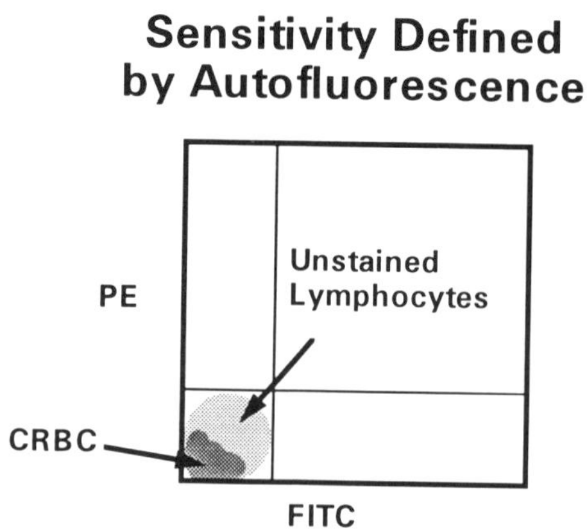

Fig. 2.10. Autofluorescence of unstained cells. All peripheral blood leukocytes have autofluorescence. Lymphocytes and progenitor cells have the least, while monocytes and granulocytes are much more autofluorescent. The sensitivity of a flow cytometer may be capable of discriminating autofluorescent lymphocytes from nonfluorescent cells and can be checked using a suspension of cells with even less autofluorescence, such as erythrocytes (e.g., chicken red blood cells [CRBC]).

The flow cytometer has inherent limitations based on the optical configuration and detectors of the system. The lower limit of detection also depends on the flow rate of the cell through the laser, the laser power, and cleanliness of the system. For most flow cytometers, the instrument is more sensitive than necessary to measure the intrinsic fluorescence (autofluorescence) of the cells being analyzed.

Unstained cells have intrinsic fluorescence that comes from the biologic components within the cell. This fluorescence, called autofluorescence, varies with cell type. Lymphocytes exhibit the least autofluorescence, while monocytes and granulocytes have the most. Progenitor cells have an autofluorescent pattern similar to that of lymphocytes.

The autofluorescence of unstained lymphocytes is equivalent to approximately 1000 fluorescein molecules. As a flow cytometer can detect much lower levels of fluorescence, the lowest limit of cellular detection is controlled by the sample itself and not by the instrument. Fresh erythrocytes have less autofluorescence than unstained lymphocytes. Osmium fixed cells have even less fluorescence than fresh erythrocytes and may be used to assess instrument sensitivity.

COMPONENTS OF A FLOW CYTOMETER

The components of a flow cytometer can be identified by their role in obtaining quantitative information about the sample. Each component can affect the quality of the data and therefore requires instrument QC.

Optics

The optics of the flow cytometer filter and focus the light used to identify the different characteristics of cells. The optical systems include both the components for manipulating the laser beam to illuminate cells and the light collection lenses, filters, and detectors used for collecting signals coming from the cells.

Multiple focal points. Optical and flow paths must be aligned and focused. The laser light must be positioned and focused onto the cell stream as shown in Figure 2.11. The forward light scatter detector must be centered on the laser beam and also focused on the

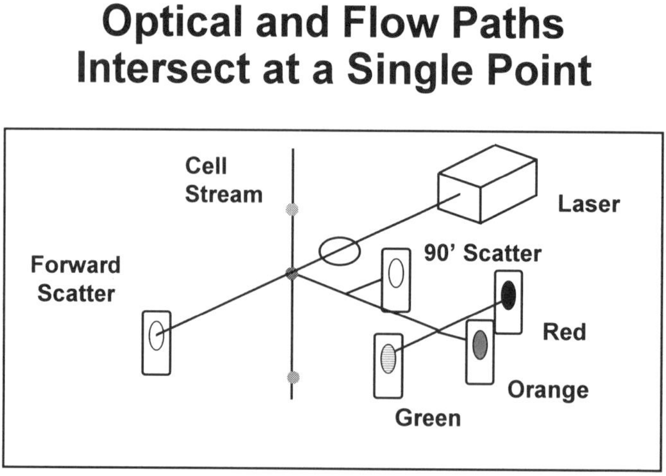

Fig. 2.11. Intersection of optical and flow paths. Cells flow in line through the focal point of the laser beam and the intersection of the 90° light scatter and fluorescence detectors.

cell stream. All detectors at right angles to the laser beam also are focused onto the cell stream.

Each cell must flow in line through the laser beam and through these multiple focal points.

Light scatter. The amount of light scattered can provide information about the size, refractive index, and granularity of each cell passing through the laser light. Cell size is relative to *diffraction* of light, refractive index is measured by *refraction* of light, and the irregularities in the cell are observed by *reflection*. Light scatter does not change the color of the light. This means that the forward and right angle light scatter detectors do not require spectral filters such as those needed by the fluorescence detectors.

The amount of light scattered by a cell changes dramatically as a function of the collection angle relative to the incident laser beam. Forward angle light scatter measures cell size by a combination of diffraction and refraction. Right angle light scatter detects internal cytoplasmic and nuclear complexity (granularity) by reflected light.

Forward light scatter. Forward light scatter is detected along an axis parallel to the incident laser beam. A small beam stop is positioned in front of the detector to block incident laser light from entering the light scatter detector. The forward light scatter detector then collects light just off-axis from the laser beam. The principle is similar to that of dark field illumination in a microscope.

1. The forward light scatter signal is affected both by the cross-sectional area of the cell and the refractive index difference between the cytoplasm and the sheath fluid.
2. The forward light scatter signal is a relative rather than an absolute measure of cell size because of an interaction between cross-sectional area and refractive index.
3. Dramatic changes in light scatter signals occur as the detector collects light closer to the laser beam. This means that small changes in laser position or beam stop size can give large changes in discrimination in cell types.

Right angle light scatter. Right angle light scatter is collected at right angles to the laser beam. This is also known as 90-degree light scatter (90°SC), side scatter (SSC, SS), or orthogonal light scatter.

1. Right angle light scatter measures the difference in granular density and composition inside the cells. The cell granules differ in their ability to reflect light into the detector.
2. There is a slight effect on right angle light scatter intensity from differences in nuclear shape and cell-surface irregularities.
3. Cells which are not spherical, such as erythrocytes, can also give rise to large right angle light scatter signals.

Fluidics

Fluidics ensure that cells flow through the instrument in single file and arrive at the laser beam in exactly the same position every time. This is accomplished by injecting the cell suspension into the center of a stream of cell-free sheath fluid. A hydrodynamic focusing effect is created that confines the cells to the center of the stream and ranks them in single file. It makes no difference whether this concentric flow is upwards or downwards relative to the laser beam.

The path that the cells can take is limited by the depth of field (the area which remains in focus) of the various optical components. This means that the cells must be confined to a given width within the stream. Any blockage, partial blockage, or bubble can change the path of the cell, thereby causing it to pass outside the focal point of one or more of the detectors of the flow cytometer. This changes the signal coming from the cell and may result in an incorrect classification of that cell because it could not be compared with others that did pass through the optical center.

Time cells are exposed to laser. Cells are only in the laser beam a short period of time (on the order of microseconds). If the cell is in the laser beam for a longer period of time, the signal coming from the cell can be increased. Thus, the longer the cell is in the laser beam, the more fluorescence it will emit, effectively increasing sensitivity of fluorescence detection. Time in the laser is related to the sheath pressure, which controls *cell velocity*. Sample pressure only affects the cell rate, the frequency of cells intersecting the laser.

Sheath pressure. Since the time that a cell is in the laser beam depends on the flow rate of the sheath fluid, changing the sheath pressure on some flow cytometers can affect the time the cell is in the path of the laser light and thereby change the signal coming from the cell. The sheath pressure may be changed for applications other than immunophenotyping, such as measuring the DNA content.

If the sheath pressure on the flow cytometer is changed, the entire standardization of the instrument must be repeated because the sensitivity of the instrument has been changed.

Sample pressure. While the sheath pressure can affect how long the cell is in the laser path, the sample pressure affects how many cells are injected into the sheath fluid. This controls the cell frequency, the number of cells analyzed per second, and the time required to collect the necessary number of events on any particular sample. The sample pressure does not affect cell velocity.

Sample Preparation

Different sample preparation techniques can cause cells to swell or to shrink. In addition, fixation can cause changes in the refractive index of the cells. This means that changing sample processing protocols can affect the resolution observed on the same instrument, (Fig. 2.12). In order to minimize variation and maximize consistency, the same sample processing procedures should be used consistently.

Specimen problems. Nonlymphocytes must be identified and accounted for if quantitative lymphocyte subsets are to be accurately

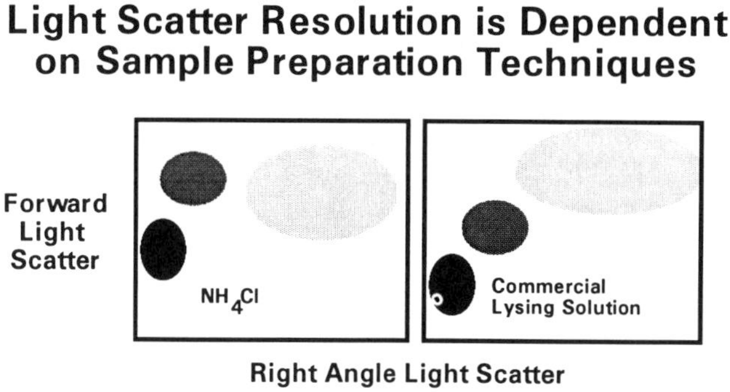

Fig. 2.12. Effect of red cell lysis method on light scatter resolution. Commercially available solutions for red cell lysis can cause the remaining white blood cells to swell or to shrink. The resulting light scatter patterns can be characteristic for each lysing procedure.

determined. Dead cells scatter less light than live cells as shown in Figure 2.13. Granulocytes that have been activated may lose their granules and begin scattering light as if they were lymphocytes. Platelet aggregates can form during sample processing and scatter light like lymphocytes. Erythrocytes from different specimens may react differently to each lysis protocol; so that some samples may contain unlysed erythrocytes. Internal QC checks must be made to identify and to correct for such specimen variability.

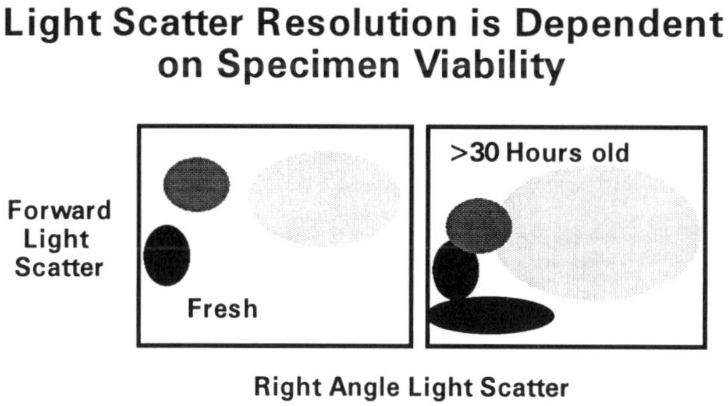

Fig. 2.13. Effect of cell viability on light scatter resolution. Peripheral blood leukocytes from old specimens will have poor light scatter resolution. All populations will scatter light less, moving the dot plots in a downward smear, making accurate population discrimination very difficult.

Differential analysis. The light scattering signal can provide a differential analysis of the cells in blood, including lymphocytes, monocytes, granulocytes (neutrophils, eosinophils, and basophils), and debris (platelets and erythrocyte ghosts). This differential must match that obtained from an independent hematology instrument if the calculation of absolute counts is to be accurate.

Removal of erythrocytes. The leukocyte differential analysis requires the removal or lysis of erythrocytes. The ratio of leukocytes to erythrocytes is approximately 1 : 1000, so that without erythrocyte removal, leukocyte subset analysis is a rare event.

Removal of platelets. Platelets must also be eliminated from the analysis since aggregated platelets can be the same size as leukocytes. This is accomplished during sample preparation by centrifugation or by discrimination using forward light scatter.

Electronics

The electronics process the signals coming from the photodetectors. This processing includes signal shaping, amplification, and conversion from analog to digital format.

Photodetectors. Two types of photodetectors are used in flow cytometers: photodiodes and photomultiplier tubes. Photodiodes are used as the forward light scatter detectors because their low sensitivity is overcome by the intensity of the laser light. These detectors have no high-voltage settings. Photomultiplier tubes (PMT), on the other hand, are high-sensitivity detectors, requiring high voltage and having variable spectral sensitivity. They are used for the right angle scatter and fluorescence detectors, which receive less signal than the forward scatter detector. Photodetectors convert the photons of light striking the detector to electrical impulses proportional in magnitude to the number of photons received. A change in the PMT high voltage changes the amplification of the detector in a nonlinear fashion.

The phototubes have variable spectral sensitivity, with the greatest sensitivity being in the blue part of the emission spectrum, rather than the red.

The high-voltage setting for each of the flow cytometer's photodetectors will differ because each detector is measuring a unique cellular characteristic. The voltages must be adjusted for the optimal detection of fluorescence and the right angle light scatter. High-voltage settings are not required for the forward light scattering photodiode.

Signal amplification. Signal amplification is accomplished by increasing the voltage applied to the PMT or by changing the gain on the amplifiers. Changing the PMT voltage effectively changes the sensitivity of the detector as it is converting the signal. Such voltage changes cause nonlinear changes in the signal output.

Changes in the amplifier gain cause changes to occur in the signal after it has been generated by the PMT. Therefore, this amplification process is relatively linear depending on the electronic components. For very careful research studies or DNA analyses, it may be necessary to calibrate the amplifiers to determine their linearity. This is usually not necessary for immunofluorescence analysis.

Signals from the photodetectors can be amplified in a linear or logarithmic fashion as shown in Figure 2.14.

Linear amplification is usually used for the forward light scatter detectors. This results in a display of the forward light scatter signal on a linear axis. The linear amplifier is affected by gain settings.

The shapes of histograms are difficult to compare when using linear amplification. This method of amplification tends to emphasize bright populations as compared with dim populations.

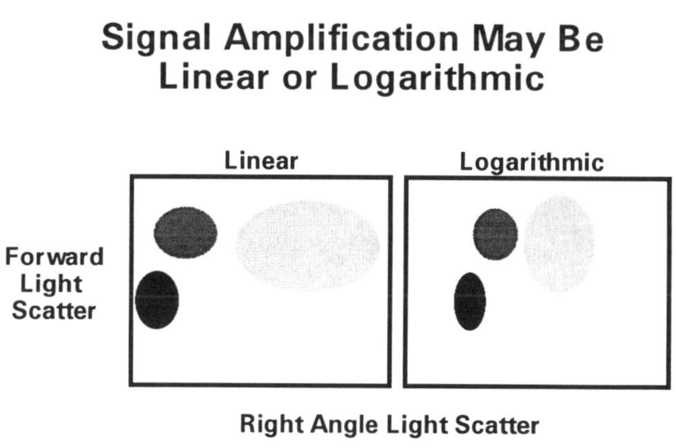

Fig. 2.14. Effect of linear or logarithmic amplification on right angle light scatter resolution. Linear amplification usually creates a much broader light scatter display and emphasizes bright populations. Logarithmic displays show all populations closer together.

Logarithmic amplification is usually used for the fluorescence channels. Logarithmic amplification does not require gain settings and is usually set to four decades displayed full scale. The four decades are required to identify both dim and bright immunofluorescence-stained cells. The right angle light scatter channel can be amplified in either the logarithmic or linear mode.

Logarithmic amplification preserves the shape of curves even though they may appear at different intensities. This amplification emphasizes dim rather than bright populations. It should be noted that zero is undefined in a logarithmic plot.

Signal digitalization. The analog signal coming to the computer must be converted to a digital signal. The process of conversion is performed by an analog-to-digital converter. For an event with a larger analog signal, a correspondingly larger number is produced.

The number of divisions or channels into which the analog signal can be subdivided determines the resolution of the instrument. Some instruments divide the analog signal into 64 different segments, while others divide the same signal into 1024 segments. The ability to resolve one intensity as being different from another is better when the scale is divided into more segments (channels).

Spectral compensation. The colors emitted from the dyes used in immunofluorescence cover a broad portion of the visible spectrum. The optical filters used to separate the emissions from these dyes pass light only in small regions of the spectrum closely matched to the peak emission of the dyes used in the immunofluorescence analysis. Some dyes emit light that passes through more than one of the instrument's filters; for example, fluorescein emits not only green but some yellow-orange light. This orange light passes through the orange optical filter and will generate a signal in this detector. Similarly, some red light from PE can pass through the red optical filter.

Correction for the spectral overlap is usually made by the electronics during data acquisition, although computer recalculation of raw data can be made with proper software. The electronic correction is made using the analog signals and subtracting a portion of the signal of one color from the other colors. Proper setting of the compensation circuit is critical to fluorescence analysis. Overcompensation counts dim cells as negative, while undercompensation identifies some negative cells as positive.

Data Processing

A computer program acquires the data, then classifies and groups it according to the four of five signals detected for each cell. When the data are collected and then stored (list mode data), it may be processed at a later date by analysis software. Displays of the data can then be generated in the form of single-parameter histograms, two-color dot plots, or contour plots.

List mode data. Each event or cell generates five separate numbers: forward light scatter, side scatter, and three colors of fluorescence. The maximum information about the sample can be stored by using list mode data storage. In list mode data storage all of the numbers are stored for each cell in sequence. The list is as long as the number of cells that are collected for that sample.

The data from list mode can then be analyzed as many times as needed, allowing changes in the analysis each time, without changing the actual data set. This preserves the maximum amount of information; however, it uses large amounts of computer storage space.

Data storage. Data storage requirements are mandated by regulatory agencies. The original report sent to the physician must be kept for a period of time consistent with state requirements, generally at least 2 years. It is advisable to maintain a hard copy of the data used to generate the report as part of the patient file. In this way questions from the physician can be answered by referring to the hard copy of the data that was used to generate the clinical report.

Storage of list mode files that were used to generate the report is preferred since the data can be reanalyzed at any time. Storage of list mode files, however, is expensive, considering the relatively few requests for reanalysis of data for standard immunophenotyping protocols. Many laboratories only maintain list mode files for 30 days after the final report is issued.

CONCLUSIONS

Understanding the components of flow cytometric immunophenotyping is essential for a laboratory developing a comprehensive quality assurance program. Quality assurance and QC must address all assumptions made in the process and must control all phases of the instrumentation and procedures.

FURTHER READINGS

Caldwell CW, Maggi J, Henry LB, Taylor HM. Fluorescence intensity as a quality control parameter in clinical flow cytometry. *Am J Clin Pathol* 1987, 88:447–456.

Gelman R, Cheng SC, Kidd P, Waxdal M, Kagan J. Assessment of the effects of instrumentation, monoclonal antibody, and fluorochrome on flow cytometric immunophenotyping: a report based on 2 years of the NIAID DAIDS flow cytometry quality assessment program. *Clin Immunol Immunopathol* 1993, 66:150–62.

Hansen W, Hoffman R, Ip S, Healey K. Light scatter as an adjunct to cellular immunofluorescence in flow cytometric systems. *J Clin Immunol* 1982, 2:32–41.

Horan P, Loken M. A practical guide to the use of flow systems. In Van Dilla M, Dean P, Laerum O, Melamed M, eds. *Flow Cytometry: Instrumentation and Data Analysis*. Academic Press, New York, 1985, pp. 260–280.

Loken M, Parks, D, Herzenberg L. Two-color immunofluorescence using a fluorescence-activated cell sorter. *J Histochem Cytochem* 1977, 25:899–907.

Loken M, Stall A. Flow cytometry as an analytical and preparative tool in immunology. *J Immunol Methods* 1982, 50:85–112.

Muirhead K, Schmitt T, Muirhead A. Determination of linear fluorescence intensities from flow cytometric data accumulated with logarithmic amplifiers. *Cytometry* 1983, 3:251–265.

Parks D, Herzenberg L. Fluorescence-activated cell sorting: theory, experimental optimization, and applications in lymphoid cell biology. *Methods Enzymol* 1984, 108:197–241.

Rabinovitch PS, Torres RM, Engle D. Simultaneous cell cycle analysis and two-color surface immunofluorescence using 7-amino-actinomycin D and single laser excitation: applications to study of cell activation and the cell cycle of murine Ly-1 B cells. *J Immunol* 1986, 136:2769–2775.

Ritchie A, Gray R, Michlem H. Right angle light scatter: a necessary parameter in flow cytofluorometric analysis of human peripheral blood mononuclear cells. *J Immunol Methods* 1983, 64:109–117.

Shapiro H. *Practical Flow Cytometry*, 2nd ed. Alan R. Liss, New York, 1988.

Smart Y, Cox J, Murphy B, Enno A, Burton R. Flow cytometric enumeration of absolute lymphocyte number in peripheral blood using two parameters of light scatter. *Cytometry* 1985, 6:172–174.

Stelzer GT, Shults KE, Wormsley SB, Loken MR. Detection of occult lymphoma cells in bone marrow aspirates by multi-dimensional flow cytometry. *Prog Clin Biol Res* 1992, 377:629–635.

Stites D, Casavant C, McHugh T, et al: Flow cytometric analysis of lympho-

cyte phenotypes in AIDS using monoclonal antibodies and simultaneous dual immunofluorescence. *Clin Immunol Immunopathol* 1986, 38:161–177.
Thompson J, Gralow J, Levy R, Miller R. The optimal application of forward and ninety-degree light scatter in flow cytometry for the gating of mononuclear cells. *Cytometry* 1985, 6:401–406.

STUDY QUESTIONS

1. Name the types of photodetectors used by the flow cytometer.
2. Name the photodetector that is characterized by low sensitivity and flat spectral response, and does not require high voltage.
3. Name the photodetector characterized by high sensitivity, high voltage requirement, and variable spectral sensitivity.
4. Name three types of flow cytometric data displays.
5. Name the type of electronic signal that is initially collected by the flow cytometer and the type of signal that is processed by the FCI software.
6. Discuss the similarities between the fluorescence microscope and the flow cytometer in adjusting the light source, focusing the optics, and selecting the appropriate optical filters.
7. Explain the difference between determining cell morphology with a fluorescence microscope versus assessing cell morphology with a flow cytometer.
8. Discuss the effects on the data of varying the intensity of the laser beam, either by changing focus of the laser or by changing the flow path relative to the laser light.
9. Describe the effect on the results of excessive cell clumps or debris in the sample.
10. Explain the relationships among assay sensitivity, cell velocity, sheath pressure, and the length of time each cell remains in the laser beam.
11. Explain the relationship between sample pressure and the flow rate.
12. Describe how diffracted, refracted, and reflected light collected by the flow cytometer provide information about cell characteristics.

13. Name two major categories of light scatter and the primary cell characteristic that each measures.
14. Draw a two-parameter light-scatter plot showing the relative shape and position of the lymphocytes, monocytes, granulocytes, and debris.
15. Describe why defining lymphocytes by light scatter alone may not be ideal.
16. Name the two most widely used fluorochromes in FCI and two major characteristics of each dye.
17. Describe why optical filters must block laser light, transmit fluorescent light, and discriminate between fluorescence emissions.
18. Explain why electronic compensation is required for the optical filters.
19. Describe how the autofluorescence of unstained cells is used to define instrument sensitivity.
20. Name the type of data storage that maintains digitized numbers for each photodetector reading for all the cells in the sample, while preserving the maximum amount of information for each data set and allowing multiple changes in analysis of the sample.

III

Quality Control of Flow Cytometer and Reagents

INTRODUCTION

The purpose of instrument quality control, which includes initial setup, alignment, and testing, is to ensure that the flow cytometer operates identically from day to day. The flow cytometer is a complex instrument with many different adjustments that can change over time; therefore it is important to provide a standard setup at the beginning of each run to ensure that the instrument is operating consistently each day. In addition, it is necessary to test the expected fluorescence staining patterns of the monoclonal antibody reagents and to design quality control specifications for performance.

INSTRUMENT QUALITY CONTROL

Proper instrument quality control must test all of the components with respect to established specifications. This includes the major components of optics, electronics setup, including photomultiplier tube (PMT) voltages, and spectral compensation. Within the optics it is important to test both light scatter detectors and fluorescence detectors.

Assumption: It is assumed that the flow cytometer has enough sensitivity to enumerate leukocyte and lymphocyte subsets.

This major assumption can be further divided into three additional assumptions:

1. It is assumed that discrimination between lymphocytes and other cells by light scatter is adequate. This means that lymphocytes can readily be distinguished from monocytes and debris, and that this resolution is optimal.
2. It is assumed that the sensitivity in the fluorescence channels allows distinction between dim and negative cells. This is important to show that the instrument is sensitive enough to perform quantitative immunophenotyping.
3. It is assumed that the instrument operates identically every time, all the time. A standard must be run to show that the instrument is set up properly at the beginning of the run. Periodic monitoring is necessary to show that the sensitivity has not changed during operation.

Monitoring instrument performance can be accomplished with standards, calibrators, or controls. The cells or particles comprising each of these suspensions are described below:

Standards

Standards are stable particles with characteristics that closely match those of the cells to be tested. The chromophores on the standard may not exactly match the fluorescent dyes used for immunophenotyping; therefore, no absolute quantitation of dye equivalents can be made for comparison with patient staining patterns.

A standard, however, can provide reproducible, fluorescent positions with respect to stained cells. Standards are used for instrument setup. The standards may be cross-referenced to other particles such as calibrators in order to define absolute fluorescence quantities for analysis. Because some standards may not have the same chromophores as those used for immunophenotyping, they *cannot* be used to set spectral compensation.

Calibrators

Calibrators are particles that have been conjugated with fluorescein, phycoerythrin, or other chromophores used in immunophenotyping. The amount of fluorescence for each of the particles has been quantified.

Mixtures of these particles with different known intensities provide a test of the linearity of the instrument. Such particles may have a limited shelf life and/or may change in their properties over a period of time.

The calibrator particles may be used to set spectral compensation as long as they are conjugated to the same fluorochromes used in the patient analysis.

Spectral compensation must, however, always be checked by using stained cells.

Controls

Controls test the entire assay procedure including reagents, instrument, and cell processing. Controls may be lyophilized cells or fresh cells from normal individuals, which are run to assess the entire assay procedure. These may not have the appropriate characteristics for standardization of the instrument. Usually an instrument is set up using a standard or calibrator particle and then process control cells are run to test the entire analytical procedure.

Requirements for Standards

In order to serve as a good standard for FCI, a particle must be similar to the cells being tested, mimic the signal intensity of stained cells, be able to test light scatter as well as fluorescence signals, and be stable for long periods of time. It is not proper to set up the instrument for immunophenotyping using a standard that is much brighter, much larger, or much smaller than lymphocytes.

Similar to cells. The standards should possess characteristics of size, granularity, and fluorescent intensity similar to those of cells and should be used at intervals during the day to continuously check the instrument sensitivity. This requires that standards be easy to obtain and relatively inexpensive so that there is little concern about using them frequently.

Test all channels. The standard must be able to test all optical channels, including forward light scatter, right angle light scatter, and fluorescence channels. The standard may be a single particle with a broad-spectrum emission or a series of particles, each standardizing a single channel.

Defined peaks. The standard must have well-defined peaks in each of the channels being tested (Fig. 3.1). The peak position can then be used to define proper instrument setup. A standard with a broad fluorescence distribution cannot be reproducibly set to a specific position. The broad fluorescence staining observed for control cells precludes them from being good instrument standards.
Stability. The standard must have stable characteristics so that it can be stored for a long period of time. In general, the standards are refrigerated as a concentrate. A working suspension is then made up daily for use on that day only. The working standard may change its characteristics after 24 hours in the diluted form; thus, it is essential to know when a working standard was made in order to minimize changes that can occur over time.

Specifications

As a part of instrument quality control, it is necessary to define the minimum specifications for proper operation of the flow cytometer. These are the minimum criteria that will provide the correct results. Minimum operating specifications may be set by a manufacturer,

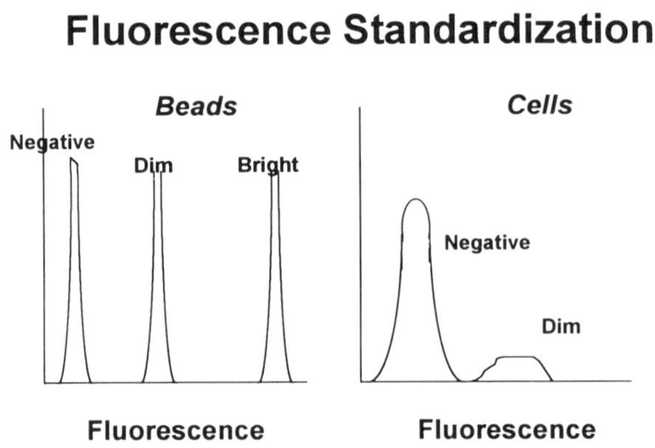

Fig. 3.1. Standardization with fluorescent beads and cells. Good instrument standards must have well-defined peaks and make it possible to distinguish bright from dim fluorescence, and nonfluorescent particles from dimly fluorescent particles. Fluorescent-stained cells are not good instrument standards because they give broad peaks and poor discrimination between negative and dimly stained cells.

or by laboratory operations staff after they have gained experience with the instrument in the flow cytometry laboratory. There are a number of considerations when defining specifications.

Who sets specifications? Manufacturer's specifications are usually defined in terms of specific standards used in instrument setup. Because standards may be unstable or have a limited shelf life, manufacturers set broad specifications that may not reflect optimal operation of the instrument for immunophenotyping. Laboratory-defined specifications may be more realistic in terms of daily fluctuations and trend analysis.

Historical use of standards. Changes in standards are easily detected when a particular instrument has been operated over a long period of time. Through long-term operation of a flow cytometer it is possible to establish routine settings and positions of standards for each instrument. Daily records of these settings and positions can then be used for long-term monitoring of the instrument.

It is important to compare the settings obtained just after the instrument has been serviced with those in the daily log sheet. If the settings indicate a more sensitive instrument after servicing, then the specifications should be upgraded to the new settings. If, however, the instrument is less sensitive, the service representative should not leave until the instrument matches or exceeds the previous sensitivity values.

INSTRUMENT SETUP

Instrument checks must be performed at the beginning of each run to show that the instrument meets minimum requirements. Each instrument must be tested for optical alignment, electronic standardization, sensitivity, and compensation.

Optical Alignment

The goal of optical alignment is to optimize the signal in each of the sensors. This is performed by using standard particles and adjusting the alignment to place the signals in the same position for each run. If it is not possible to maximize all parameters, it may be necessary to have a service representative adjust the instrument.

It should be noted that alignment on some flow cytometers can only be performed by trained service personnel.

Optical alignment must be performed and/or tested for all of the components, including the three fluorescence channels and two light scatter channels. The laser focus and positioning must be tested, as well as the alignment of the cell path through the single focal point of all the optical channels.

During alignment it is necessary to adjust the detectors, maximizing the signals generated by standard particles. Proper alignment is attained when all of the components are optimized. Small changes in one adjustment can affect other parameters; so that optimizing the signal in one component may reduce the signal in another.

Optical focus. All optical components of the flow cytometer converge on a single point. The laser light must be positioned properly to intersect the forward light scatter detector correctly. It also must be focused onto the intersection of the stream with the other focal points.

The forward light scatter detector must be focused on the intersection of the cell stream and the laser beam. The detectors that are orthogonal to the laser beam must be focused on the central point.

The cells must also be hydrodynamically focused to pass through the intersection of the optics. The cells must be confined to a relatively small internal stream. This keeps the stream width relatively narrow compared to the depth of field of the individual focal paths.

Dirty optics and fluidics. Dirty optics, resulting from dust and dirt collecting on lenses and on the flow cell, can minimize the effective laser power and thereby reduce the signals in each of the channels. It is necessary to keep the flow cell and fluidics clean, since partial plugs and bubbles can cause cells to flow through different parts of the laser beam. In addition, dirty flow cells can cause an increase in the background fluorescence.

Electronics Standardization

Electronics standardization plays an important role in obtaining reproducible data. By changing the PMT voltages, it is possible to place a signal almost anywhere on the display. Proper instrument setup and voltage adjustment place the standard particles in exactly the same position day after day so that the data obtained over time look identical.

Misalignment or suboptimal optical setup may be hidden by increasing PMT voltages or gain settings. It is important to know the electronic settings over a period of time to identify any consistent drift in one or more of the components.

Electronics standardization is performed after optimizing the alignment of the laser. Two approaches to electronic setup are listed below.

1. One method uses exactly the same electronic settings from one day to the next and requires *monitoring where the standards appear* in each of the five channels.
2. An alternate method involves adjusting the electronics to place the standards in exactly the same channel every time and then monitoring *the changes in electronic settings*. These adjustments are usually small compared to the high voltage applied to the PMT on the fluorescence detectors. Similarly minor adjustments can be made on the gain settings for the linear amplified channels (forward and right angle light scatter).

In either case, all electronic settings and channel numbers should be documented and used to monitor instrument consistency.

Unstained cells. The first time an instrument is run for immunophenotyping, the settings should be established with unstained cells. The forward and right angle light scatter detectors are set so that lymphocytes are observed approximately half way up the forward angle light scatter axis, while granulocytes are observed just on-scale on the right angle light scatter axis (Fig. 3.2).

Once the scale is set, three populations of cells can be identified morphologically by their positions in the light scatter plot. It does not matter which light scatter detector is shown on the y-axis or the x-axis; the results are identical although the data give different patterns (Fig. 3.3).

The fluorescence detectors are set with logarithmic amplification. The PMT high voltages should be set so that the unstained cells

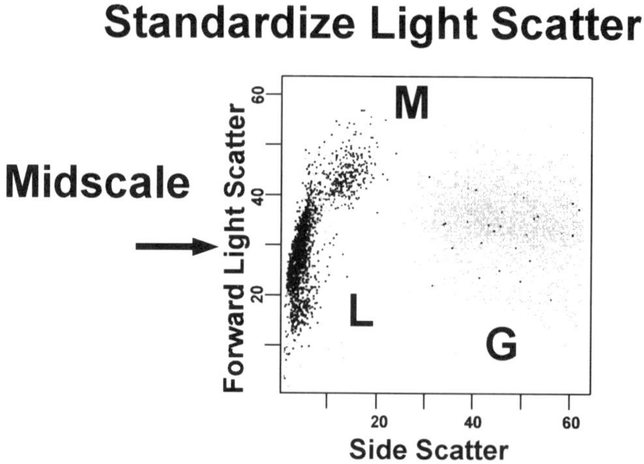

Fig. 3.2. Positioning cell populations with light scatter. Lymphocytes should be placed at midrange of the forward light scatter axis. Granulocytes should be brought on-scale on the side scatter axis.

are displayed in the lowest decade in each of the fluorescence channels (Fig. 3.4).

Standard particle. Once the electronics have been standardized, standard particles selected by the laboratory can then be run and the position in all five channels recorded. It is necessary to record the position of the standards and the electronic settings each time the instrument is set up. It is also necessary to ensure that unstained cells appear in the same place every time.

Once the initial position of the standard has been defined, the instrument can be set up by placing that standard in the exact same spot for each subsequent run.

It should be noted that unstained lymphocytes from various donors may vary slightly in their position. For this reason one uses a standard particle to set up the instrument and then the unstained or autofluorescent cells to check the setup.

Sensitivity Specification

It is important to establish the minimum instrument sensitivity for proper operation. This defines the conditions under which the re-

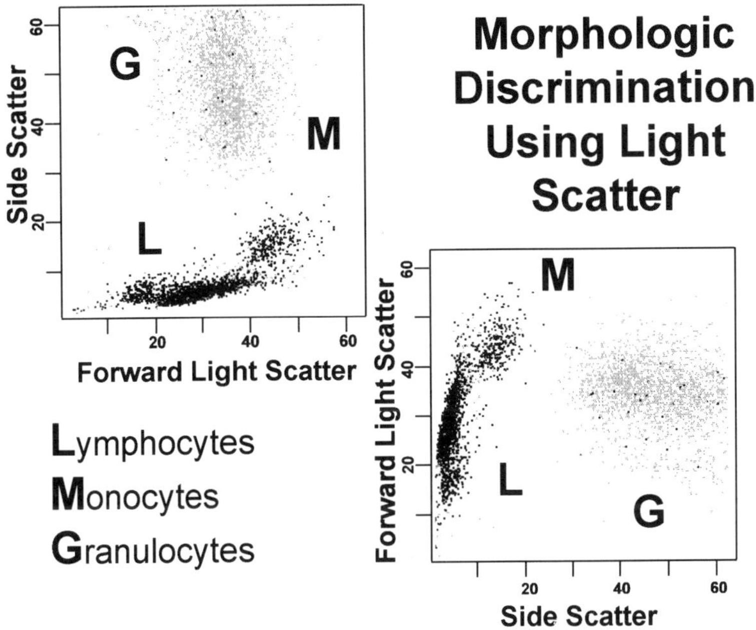

Fig. 3.3. Identifying three populations of cells by light scatter pattern. Lymphocytes, monocytes, and granulocytes can be distinguished by light scatter patterns, regardless of which light scatter parameter is displayed on the y-axis.

sults of the tests are considered valid. If the instrument is not sensitive enough, the results may be questionable.

It is important to show that the instrument is sensitive enough to perform FCI analyses. The sensitivity of the instrument is assessed by measuring the separation between at least two standards (Fig. 3.5). If the instrument is stable and the sensitivity remains constant, the distance separating two standard particles will remain constant (e.g., log channels separation of unstained cells and positive standard in each detector). If the sensitivity is decreasing, then the distance between the two standards will decrease.

Cross-standardization. Cross-standardization should be performed by comparing one standard with another. This must be done between primary and secondary standards or when new lots of the primary standard are purchased.

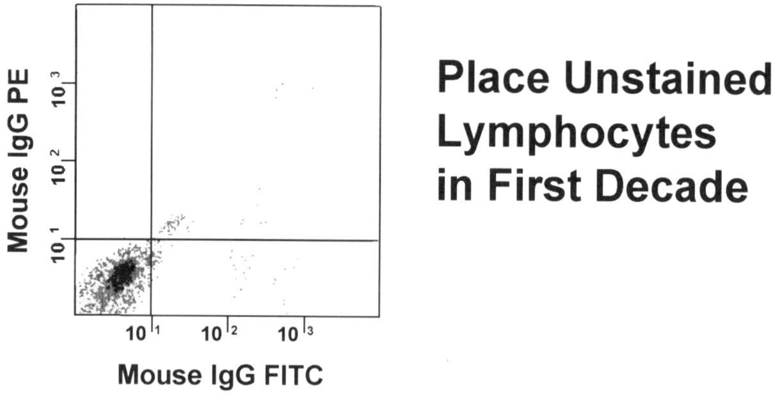

Fig. 3.4. Positioning the negative cell population. Unstained lymphocytes, incubated with irrelevant mouse immunoglobulins, should be run and placed in the first decade of each fluorescence detector. This population serves as a control for the correct position of lymphocyte autofluorescence.

The instrument is first set up using the primary standard and is shown to meet minimum specifications. The second standard is then run and its location relative to the first standard is noted and recorded.

When daily records of the position and settings for the primary standard are maintained along with weekly or monthly records of

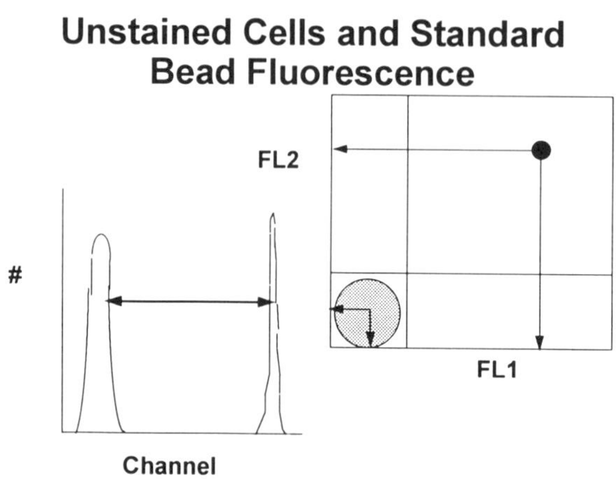

Fig. 3.5. Measuring fluorescence sensitivity. Instrument sensitivity is assessed by measuring the separation of mean channel fluorescence intensity of two standards. A stable flow cytometer will show consistency in these measurements.

the positions of the secondary standard, it is possible to use the secondary standard in place of the primary standard at any time.

Secondary standards (backup). Secondary standards are used to assess the stability of the primary standard and to provide backup for instrument setup. They must be cross-standardized with the primary standard material. If the relationship between the primary and secondary standard has been defined, it is possible to use the secondary standard for instrument setup if, for some reason, the primary standard is unavailable or unusable.

Examples of primary and secondary standards include commercial beads (primary standard), calibration beads (secondary standard), and glutaraldehyde-fixed chicken erythrocytes (secondary standard).

Quick check of instrument sensitivity. It is possible to use a single standard as one of the reference points and unstained normal cells as the second reference point. The autofluorescence of unstained cells may vary from day to day; therefore, it is necessary to use an average value. The instrument has adequate sensitivity if it can distinguish between the autofluorescence of unstained cells and a dimmer particle such as OsO_4-fixed chicken erythrocytes or fresh erythrocytes.

The sensitivity of the test is often limited by the specimen or sample tube being analyzed rather than by the sensitivity of the instrument. Therefore, before troubleshooting problem specimens, it is necessary to first demonstrate that the flow cytometer is performing adequately.

Spectral Compensation

Spectral compensation must be made since the optical filters do not ideally match the fluorescence emission spectra. Spectral compensation should be properly set so that fluorescein- or phycoerythrin-labeled cells will not be detected in the channel for the other fluorochrome. When three-color analysis is being performed, all three fluorescence channels must be set for minimal spectral overlap. The emission spectra of the standard used to set compensation must match exactly with the immunofluorescence dyes, or the compensation network will be set improperly. The compensation network may be set with beads, but must *always* be checked using stained cells.

Different immunofluorescence conjugates of the same dye may not have identical spectra. During the manufacture of fluorescein-conjugated monoclonal antibodies, the number of fluorescein molecules attached to any antibody may vary. The more dye molecules that are attached to the antibody, the more likely it is that there are dye-to-dye interactions that may shift the emission spectra. An antibody with a fluorescein-to-protein ratio of 2 : 1 has a slightly different emission spectra than one with a fluorescein-to-protein ratio of 6 : 1.

Dye-to-dye interactions are often seen under conditions where antibodies are aggregated or when fluorescein antibodies are used to stain cells with extremely high antigen density. The microenvironment, pH, or solvent interactions may also change the emission spectra; therefore, fluorescence compensation should be checked with each new lot of antibody used.

Fluorescence compensation is based on the relative sensitivity of the detectors analyzing the immunofluorescence. Changes in PMT voltages effectively change the sensitivity of the detector; therefore, whenever the voltage is changed on a fluorescence channel, it is necessary to reset the compensation network. Similarly, changes in optical filters change the sensitivity of the photodetectors. If an optical filter is changed, it is necessary to reestablish proper compensation.

Setting spectral compensation. Spectral compensation can be set with test particles or with stained cells. It is important that, whichever approach is taken, the fluorochromes for setting compensation be the same as those in the stained patient sample.

Test particles. The spectral compensation network for two-color immunophenotyping can be properly set using standard particles that have fluorescein and phycoerythrin conjugated to them. Three-color analyses are more complicated, but the compensation network could still be set using beads, as long as the beads contain the same dyes as the stained cells.

Three particles are required for the initial setting of the two-color compensation network: fluorescein-labeled beads, phycoerythrin-labeled beads, and unstained beads. These particles can be mixed or run sequentially.

The unstained particles should be set to the lower left-hand corner of a display of fluorescein versus phycoerythrin fluorescence. They should have the same low level of intensity as the autofluorescent (unstained) cells. This step may be accomplished during the instrument standardization.

The beads with only fluorescein on them are then run and the fluorescence compensation network adjusted so that these beads are detected only in the fluorescein channel. These fluorescein-only beads should not be seen by the phycoerythrin detector and, therefore, should have the same mean intensity as the unstained beads in the phycoerythrin channel.

Beads labeled with phycoerythrin-only should be detected only in the phycoerythrin channel and not by the fluorescein detector. Proper spectral compensation should ensure that phycoerythrin-only stained beads appear indistinguishable from the unlabeled beads when assessed from the fluorescein channel. The results from a mixture of labeled and unlabeled beads is shown in Figure 3.6.

The amount of signal subtraction required for proper spectral compensation is dependent on the PMT voltage settings. Changing the PMT voltage settings requires changing the spectral compensation network. In normal instrument setup, the subtraction of the

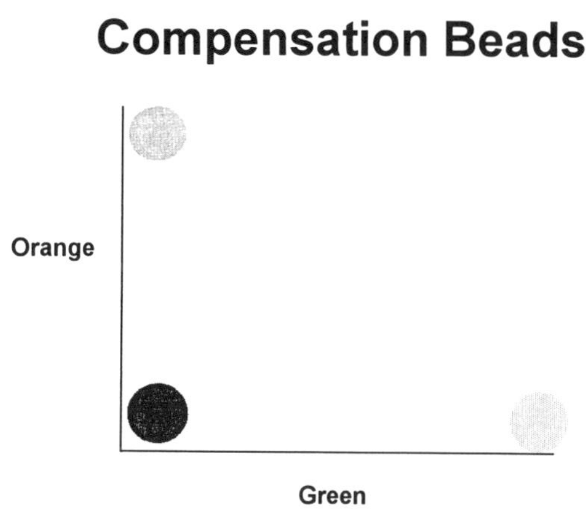

Fig. 3.6. Beads for setting the compensation network. Three beads are used for setting two-color compensation: an unstained population, and two stained populations (for example, orange and green), each bearing one of the fluorochromes, FL1 and FL2. In a proper compensation setup the FL1 beads should show the same mean channel fluorescence in the FL2 channel as the negative population, and a similar situation should obtain mean channel fluorescence for the FL2 beads in the FL1 channel equivalent to the negative population.

fluorescein signal from the phycoerythrin signal is usually greater than vice versa.

It is always necessary to use fluorescent-stained cells to check the compensation network if it was set using labeled beads. Lymphocytes stained with FITC and PE-labeled monoclonal antibodies that recognize separate cell populations should be used. The stained cells should have signals as intense as the brightest to be expected in the subsequent clinical tests.

Using cells for setting compensation. Setting spectral compensation using a preparation of stained cells is similar to setting it with labeled beads; however, the antibodies used should be selected for broad antigen expression, giving both dim and bright populations. The cells should be stained with each color monoclonal antibody separately.

The cells to be used for spectral compensation should be of the same type as those subsequently being tested. For example, if lymphocyte subset analysis is to be performed, the spectral compensation should be set up with lymphocytes. If other types of cells are used (e.g., leukemic cells or a cell line), lymphocytes may not be appropriate for establishing spectral compensation. Lymphocytes are appropriate target cells, however, for the compensation setup for CD34 enumeration.

Examples of reagent combinations for compensation setup are:

A. Non-overlapping cell populations
   1. Unstained or unlabeled cells
   2. FITC-stained cells and PE-stained cells (CD3 FITC, CD19 PE)
B. Separately labeled cell populations
   1. Unstained or unlabeled cells
   2. FITC-stained cells (CD8 FITC)
   3. Phycoerythrin-stained cells (CD8 PE)

In both examples, the unstained cells (lymphocytes) should be positioned in the lower left hand corner on a fluorescence display of phycoerythrin versus fluorescein. The FITC-labeled (green) cells and the PE-labeled (orange) cells should each be set so that imaginary lines intersecting the peaks of green-only or orange-only fluorescence and the unstained cells are parallel to each axis.

It is convenient to use the same two-color antibody reagent as that to be used in the subsequent clinical tests (CD3 FITC, CD19 PE); however, the best antibody for setting spectral compensation is one that displays a broad range of antigen intensity distribution, such as CD8. Using two aliquots of cells, each stained separately with CD8 FITC or CD8 PE, is optimal for setting compensation. The CD8 antibody stains cells both brightly and dimly and permits spectral compensation to be checked over the entire range of staining intensities for both fluorescence detectors.

When using cells stained only with CD8 FITC, the FITC-labeled cells should be positioned so that they are detected only in the FITC channel and appear identical to unstained cells when observed in the PE channel. Similarly, the cells stained only with CD8 PE should be detected only in the phycoerythrin channel and observed as unstained cells in the fluorescein channel.

In a similar manner, spectral compensation can be set for three-color analysis. Three tubes of cells should be stained with antibodies labeled with FITC, PE, or a third-color reagent. Each tube should be run in sequence to set compensation for that color. The FITC-labeled cells should not be observed in either the PE or the third-color red detectors. Likewise, the PE-labeled cells should only be identified by the PE detector and not by the FITC or red detectors. The third color, red-labeled cells, should not be observed in the FITC or PE detectors. The compensation for tandem conjugates must be set with each antibody used in the subsequent tests, since differences in efficiency of energy transfer can occur between different lots of reagents and this can result in differences in spectra for different lots.

Proper spectral compensation. Proper setting of the two-color compensation network is best demonstrated by examining results such as those shown in Figure 3.7. The data were obtained from a mixture of cells that were (1) unstained, (2) stained with CD8 FITC only, and (3) stained with CD8 PE only.

When no compensation is set, the brightly stained fluorescein-labeled cells can be detected in the phycoerythrin channel. Similarly the phycoerythrin-labeled cells can be detected in the fluorescein channel.

When the compensation network is properly adjusted, the fluorescein-labeled cells have the same fluorescence as unlabeled

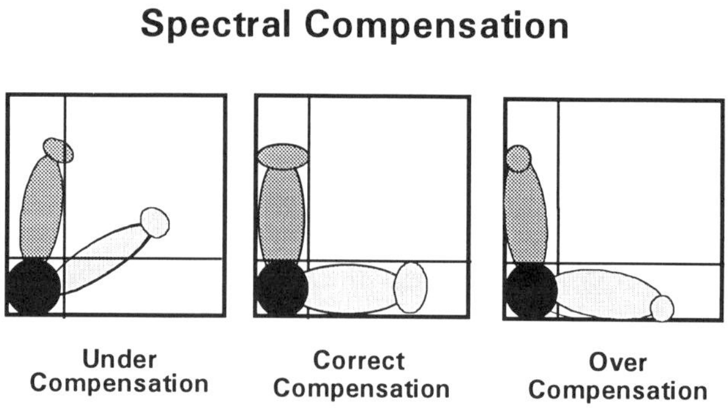

Fig. 3.7. Patterns of spectral compensation. Undercompensation allows spillover of one fluorochrome into the other detector. Overcompensation subtracts too much signal.

cells when observed from the phycoerythrin channel. Phycoerythrin-labeled cells have the same fluorescence as unlabeled cells when observed from the fluorescein channel. The populations of cells should form a right angle.

Setting the compensation network too high (overcompensation) sets the fluorescein-labeled and phycoerythrin-labeled cells off scale. The mean readings in channels for the positive cells then are lower than for the unstained cells when observed from the opposite channel. The plot for three populations of cells forms an obtuse angle when the circuit is overcompensated.

Compensation checks. It is advisable to assess the compensation settings during the normal operation of the instrument. Using the recommended panel of reagents, it is possible to check compensation on each specimen.

The position of the unlabeled cells in each sample should be identical (i.e., congruent). This position can be checked in many samples stained for two- and three-color analyses.

The position of the single-color positive cells (from any two-color pair) should form a right angle relative to the unstained cells. The use of CD3, CD19-stained cells also provides a compensation check for each specimen.

Even samples that contain double-labeled cells can be used to check proper compensation. Peripheral blood stained with CD3 FITC and CD4 PE shows three distinct populations, which appear at right angles when compensation is set correctly. With this combination the angle is opposite to that observed for cells stained with CD3 FITC and CD19 PE.

QUALIFYING THE FLOW CYTOMETER

Prior to every run, the flow cytometer should be qualified to prove that the instrument is sensitive enough to perform the test, and that is has been standardized properly for that particular run.

At the beginning of every run, a standard operating procedure should indicate that the instrument has gone through proper alignment, electronic setup, spectral compensation, and assessment of minimal sensitivity. The setup procedures should include a dated form in which the technologist acknowledges that the proper procedure was followed.

Prior to every run it is necessay to demonstrate that the instrument performs up to the specifications required to do the test. This includes meeting the necessary requirement for separation of negative from positive particles and/or the ability to resolve unstained cells from dimly stained particles.

It is necessary to keep daily records of instrument performance both to fulfill regulatory and legal requirements, and to facilitate troubleshooting instrument problems.

Daily Records of Instrument Settings

Proper record-keeping permits detection of (1) a slow drift of one or more of the detector channels, (2) a dramatic change in instrument operation, or (3) subtle changes that can assist a service technician in identifying a problem with the instrument and/or the standard particles.

It is important to maintain a record of instrument settings to check instrument performance over time. Instrument readings can be plotted versus time in a typical quality control chart (Levy-Jennings plot). This documentation can be used for either approach to electronics standardization: using the same PMT settings and monitoring channel position (Fig. 3.8), or placing the standard in the same channel each time and monitoring the PMT setting (Fig. 3.9).

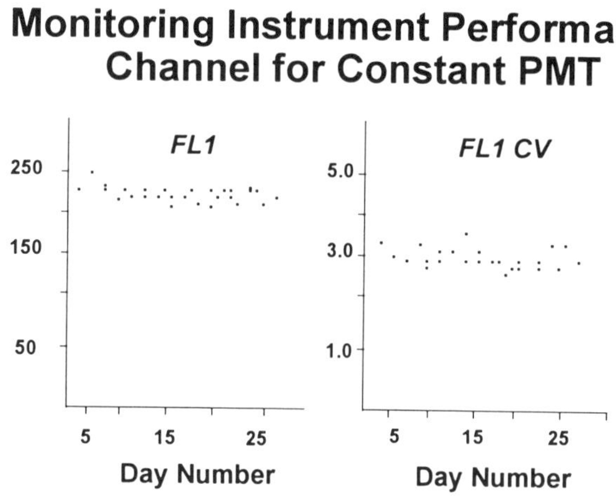

Fig. 3.8. Electronic standardization by monitoring channel position for a constant PMT setting. Levy-Jennings plots can be created to monitor instrument performance, recording both fluorescence channel number and peak CV for each fluorescence detector.

The daily record of instrument settings should include clearly defined minimum operating specifications to help the technologist identify when the instrument is not operating properly. Any corrective action can be documented appropriately.

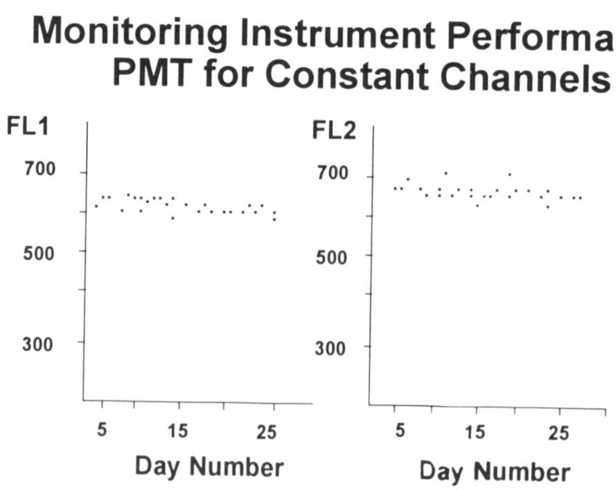

Fig. 3.9. Electronic standardization by monitoring PMT voltages for a constant fluorescence channel position. Levy-Jennings plots can be used to record changes in instrument performance for each fluorescence PMT.

The Levy-Jennings plot can also be used to aid in the cross-calibration of the lots of standards. It can also indicate when the instrument has been calibrated versus standardized, and can be used to help in checking the performance of the secondary standards.

REAGENT QUALITY CONTROL

The critical reagent for all types of immunophenotyping is the monoclonal antibody cocktail used to define subtypes of cells. Whether monoclonals are added individually or premixed, the characteristics of this reagent determine the quality of the immunophenotyping result.

Monoclonal Antibodies for Lymphocyte Analysis

A set of eight antibodies has been useful in identifying subpopulations of lymphocytes and in providing complete quality assurance checks for subset quantitation. This reagent panel has been recommended by the Centers for Disease Control and Prevention for accurate enumeration of CD4+ T lymphocytes.

Two antibodies are used to identify leukocytes (CD45) and monocytes (CD14). T lymphocytes are identified using CD3, and subsets of T cells can be identified using CD4 and CD8. B lymphocytes can be identified using CD19. Natural killer (NK) lymphocytes must be identified using a combination of reagents (CD3, CD16 and/or CD56) because no antibody has been identified that reacts solely with the NK subset of lymphocytes.

Widespread agreement on which reagents should be used to quantify CD34 cells has not been reached. The key antibody in all progenitor enumeration protocols is CD34. Usually this is matched with one or more antibodies identifying leukocytes (CD45) or myeloid cells.

CD nomenclature. The CD (clusters of differentiation) nomenclature was developed to standardize and to compare monoclonal antibodies obtained from different sources.

Monoclonal antibodies against different human leukocyte antigens were submitted for study in five international leukocyte typing workshops. These antibodies were obtained in large enough quantities to be distributed to many laboratories for a comparative analysis. The objective was to identify antibodies that had similar or identical characteristics.

The antibodies were examined based on their reactivity patterns with several cell types. These included normal blood and bone marrow leukocytes, resting and activated lymphocytes, various cell lines, and leukemia specimens. Antibodies were tested for reactivity patterns in tissue, using immunohistochemistry procedures. The antibodies were also used to identify the biochemical nature of the antigen recognized by the antibody. Although most of the antibodies were shown to identify glycoproteins with well-defined molecular weights, some of the antibodies were shown to react with carbohydrate groups or with lipids. The antibodies were also tested for their effect on various functional responses.

CD grouping. The data from all these studies were analyzed statistically to show similarities between the reactivity patterns of antibodies. The antibodies were then assigned to groups with similar characteristics and given a CD number. The creation of a new CD group required that two antibodies with similar antigen reactivity be submitted from independent laboratories.

Differences with CD groups. Antibodies assigned the same CD number may not be completely identical with regard to antigenic specificity, binding strength, and isotype.

Epitopes. Monoclonal antibodies react with epitopes, small antigenic regions on a large antigen. Assignment to the same CD number was made based on reactivity with the antigen, not with the epitope. Therefore, the recognition of epitopes may be slightly different than the recognition of antigens. Although two antibodies may react with the same antigen, they may not cross-block the binding of each other's epitope. The antibodies may also have quite different functional activities, since different epitopes of the same antigen may function differently.

Avidity. Antibodies also differ in their strength of binding to the antigen (avidity). This difference in binding strength can affect time of staining and nonspecific binding.

Isotype. Antibodies assigned the same CD number may also differ in isotype (heavy chain of the immunoglobulin). It is well known that different isotypes have quite different nonspecific binding characteristics.

Antigenic composition. Many antigens are glycoproteins. The addition or loss of a sugar moiety may change an antigen enough that some antibodies will fail to recognize the epitope.

When changing reagent suppliers, it is important to check antibody specificity to ensure that the patterns of reactivity are similar to those used previously by the laboratory. Failing to characterize different reactivities in monoclonal reagents can result in the release of clinical reports that no longer reflect the laboratory's reference range.

> **Assumption: It is assumed that monoclonal antibody reactivity is cell-type specific, unambiguous, and reproducible.**

Quality Assurance procedures must be adopted to ensure the validity of the following assumptions:

1. It is assumed that the antibody binding occurs only through the binding site so there is no nonspecific reactivity.
2. It is often assumed that a particular antibody reacts only with the cell of interest and has no reactivity with any other types or lineages of cells.
3. It is assumed that there is no variability in the manufacture of monoclonal antibodies either in suppliers or lot-to-lot variations in production.

Titering Antibodies

Immunofluorescence reagents are usually titered after they are received in the laboratory before they are used for routine testing of specimens. Reagents obtained from commerical sources should be used according to the manufacturer's recommendations. A titration, however, ensures that proper quality control has been followed to minimize lot-to-lot variation.

The proper concentration of reagents to be used in each test can be determined by a titration curve. This is performed by using two-fold (doubling) dilutions of test reagents and the standard number of 10^6 cells in 100 μL of buffer for each test. The same volume of reagent should be used at each dilution point. The peak intensity for the target cells is determined at each dilution and the fluorescence intensity is plotted versus dilution (Fig. 3.10).

As the titer point is reached, saturation of the specific antigen binding sites maximizes the intensity of the positive population before nonspecific binding decreases the separation between the positive and negative populations (Fig. 3.11).

The correct concentration of reagent depends on the procedure that is used to process the cells. For procedures in which a wash step is used prior to cell fixation, the correct dilution to choose is one dilution less than that which gives the maximum intensity plateau. Background fluorescence should be no greater than that of the isotype control for this optimal dilution.

For those procedures without a wash step before cell fixation, the optimal dilution is one dilution greater than that which gives the maximum intensity plateau. Higher concentrations of antibodies cause increased nonspecific binding when excess antibody is not washed away before fixation.

It is also possible to determine the maximum number of cells that a given concentration of antibody can stain. Doubling dilutions are used, starting from 8×10^6 cells in 100 μL, and the fluorescence

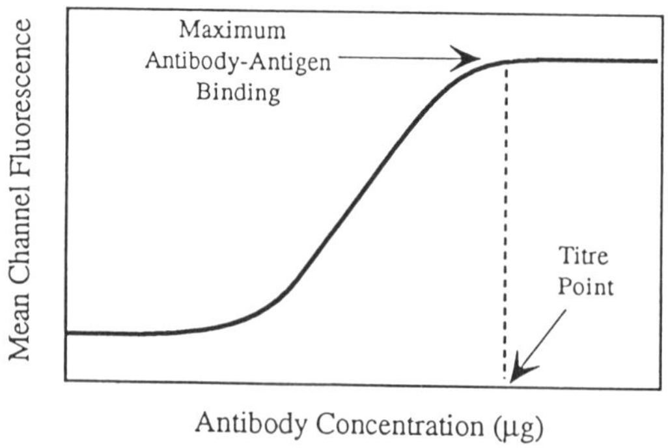

Fig. 3.10. Titration of fluorescent-labeled antibodies. Dilutions of monoclonal antibodies are tested for binding to a standard number of target cells in a standard volume. The titer point depends on immunophenotyping methodology, but is usually one dilution from the plateau position of mean channel fluorescence.

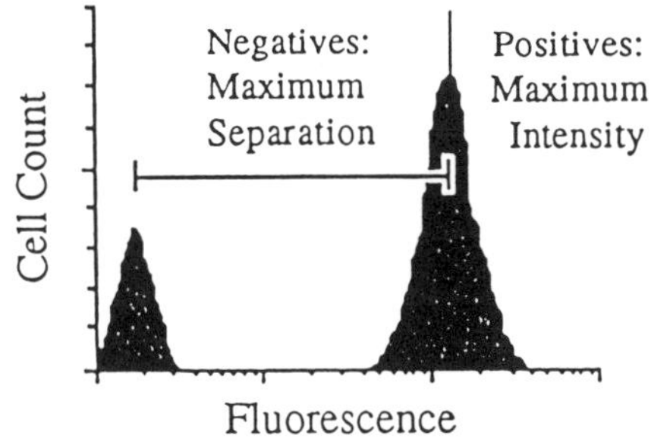

Fig. 3.11. Checking nonspecific binding at the titer point. The proper dilution of monoclonal antibody will not only achieve the highest intensity staining, it will also maintain the maximum separation between positive and negative cell populations.

intensity is noted. There is usually little change in intensity when the cell numbers are changed over a wide range.

Premixed Two-Color Reagents

Antibody reagents are provided commercially in a premixed two-color form or as single-color antibodies. Premixed antibodies have both reagents in the same bottle. This saves time and reduces the possibility of mistakes during the staining process. Optimizing the formulation of each antibody mixture to provide the best staining performance requires a checkerboard titration. For example:

1. 1.0 of A and 0.5 of B
2. 0.5 of A and 0.5 of B
3. 0.5 of A and 1.0 of B, etc.

This time-consuming step is performed by the reagent manufacturer.

Individual antibodies can also be used for two-color fluorescence staining. Often, these antibodies are sold as tissue-staining reagents, as well as reagents for FCI. As a result, the single-color reagents may be bottled at a higher antibody concentration in order to provide adequate performance on tissues exhibiting high antigen concentrations. Some laboratories may choose to use the higher concentration single-color reagents and dilute them. In this case, it is important to consider not only the proper concentration of antibody (titer) to use with each patient specimen, but also to assess potential interactions between antibodies labeled with FITC, PE, and third color.

Usually the PE-conjugated antibody is added first, followed by the FITC-conjugated antibody. A wash step may or may not be used between the addition of the antibodies. Using a wash step after staining with each reagent helps to avoid possible problems from interactions between antibody reagents. When a wash step is not used after staining with each reagent, it is necessary to test for possible interactions between antibodies. This requires that a checkerboard titration be performed to determine the optimal concentration for each antibody reagent. This is the same type of titration that is performed by manufacturers of two-color premixed antibodies in order to optimize FCI performance of the reagents.

Lot-to-Lot Testing

Comparisons should be made between lots whenever new antibodies are received. These comparisons can be carried out much faster using antibodies premixed by manufacturers. Generally, patient specimens are set up with both old and new lots of reagent and the results compared. The number of specimens required may be specified by the laboratory, but generally three to five patients are tested over 2–3 days.

Fluorescence intensity. The fluorescence intensities of the old and new lots should be identical. The titration curves should be identical when tested on the same cells. For commercial reagents, it is recommended that the initial titration begin with the recommended concentration, followed by two-fold dilutions. The same procedure can be used to validate outdated reagents. It should be noted that "super bright" reagents may have aggregated antibodies in them that can affect both specific and nonspecific staining, as well as spectral compensation.

Percentage stained. The percentage of stained lymphocytes for the two lots should be the same, within the laboratory's expected variation.

Nonspecific staining. Nonspecific staining should be the same for both old and new lots.

Agglutination. Check the forward and right angle light scattering to see if either the old or new antibodies agglutinate cells, creating very large fluorescent masses.
Compensation. Compare the compensation needed to correct spectral overlap for the old and new reagents. This is especially important with the FITC-conjugated antibodies. This should also be performed with antibody preparations that are "super bright."

Target Cells for Quality Control

Quality control of monoclonal antibody reagents is important to ensure accurate lymphocyte enumeration. Prior to use in a clinical setting, the antibodies should always be tested against target cells that express the antigens of interest. A number of different target cells can be used to test the reagents; however, it is also necessary to use fresh whole blood specimens that are stained and lysed according to the test procedure as a final QC check.
Separated lymphocytes. Separated lymphocytes isolated from a normal individual can be used for the target cell population. A titration of antibody can be made using lymphocytes separated on ficoll hypaque. In this way a large number of cells can be harvested to test a large panel of reagents. Final evaluation of the reagents should be made using fresh normal blood that is stained, lysed, and otherwise prepared by the same protocols that are used in routine analysis.
Cryopreserved lymphocytes. Cryopreserved lymphocytes can also be used for initial testing of reagents. Cryopreservation allows the collection and storage of large numbers of identical cells in order to compare lot-to-lot variations over periods of time. Freezing aliquots of a PBSC harvest is one method for providing QC for CD34 staining. This requires a cell freezer and liquid nitrogen tanks. Final evaluations of the reagent must be performed on fresh, normal blood using the procedure used in routine clinical evaluation.
Cell lines. Maintaining cell lines requires liquid nitrogen tanks for storage, and biosafety cabinets and incubators for growing the cells. Cell lines can be used for the initial testing of reagents as well as phasing in new lots of antibodies. These cells grown in tissue culture can provide extremely large numbers of target cells for the titration. Several cell lines may be needed in order to test all of the various

reagents. For CD34 titration, it is recommended that KGla cells be mixed with peripheral blood cells to a final concentration of 1–2%, simulating the proportion of PBSC. This approach allows for assessment of both positive and nonspecific staining.

It is important to understand that the antigenic expression on cell lines may change over long periods of time in culture. Therefore, a large supply of cells should be cryopreserved and stored in liquid nitrogen. Preparing a large supply of cells and freezing them into many tubes for later use helps to prevent the problem of variation in antigen expression with multiple passages. Freshly thawed cells can then be used to compare lot-to-lot variations in the antibody reagents. Again, final evaluation of reagents must be performed with fresh whole blood using routine procedures.

Lyophilized cells. Lyophilized cells have also been used for testing reagents. These have the advantage of being stable for long periods of time without the requirement for additional equipment, such as liquid nitrogen tanks for preserving cells or incubators for growing cells. Lyophilized cells can be used to compare lot-to-lot variations. Final evaluation, however, must be performed with fresh normal blood using standard processing procedures.

CONCLUSIONS

Quality control of the flow cytometer requires standardizing instrument setup and documenting the qualification of the instrument. Reagent quality control requires that monoclonal antibodies be characterized and that performance specifications be verified for each lot of reagents. Both instrument and reagent quality control performance are required for an effective quality assurance program.

FURTHER READINGS

Edwards B, Shopp G. Efficient use of monoclonal antibodies for immunofluorescence. *Cytometry* 1989, 10:94–99.

Gelman R, Cheng SC, Kidd P, Waxdal M, Kagan J. Assessment of the effects of instrumentation, monoclonal antibody, and fluorochrome on flow cytometric immunophenotyping: a report based on 2 years of the NIAID DAIDS flow cytometry quality assessment program. *Clin Immunol Immunopathol* 1993, 66:150–162.

Horan P, Loken M. A practical guide to the use of flow systems. In Van Dilla M, Dean P, Laerum O, Melamed M, eds. *Flow Cytometry: Instrumentation and Data Analysis*. Academic Press, New York, 1985, pp 260–280.

Jackson A. Basic phenotyping of lymphocytes: selection and testing of reagents and interpretation of data. *Clinical Immunology Newsletter* 1990, 10:43–55.

Loken M, Parks D, Herzenberg L. Two-color immunofluorescence using a fluorescence-activated cell sorter. *J Histochem Cytochem* 1977, 25:899–907.

Loken M, Stall A. Flow cytometry as an analytical and preparative tool in immunology. *J Immunol Methods* 1982, 50:85–112.

McCoy JP Jr, Carey JL, Krause JR. Quality control in flow cytometry for diagnostic pathology, I. Cell surface phenotyping and general laboratory procedures. *Am J Clin Pathol* 1990, 93(Suppl 1):S27–37.

National Committee for Clinical Laboratory Standards.*Clinical applications of flow cytometry: quality assurance and immunophenotyping of peripheral blood lymphocytes*. (H42-T) National Committee for Clinical Laboratory Standards, Villanova, Pa, 1992.

Parks D, Herzenberg L. Fluorescence-activated cell sorting: theory, experimental optimization, and applications in lymphoid cell biology. *Methods Enzymol* 1984, 108:197–241.

Shapiro H. *Practical Flow Cytometry,* 2nd ed. Alan R. Liss, New York, 1988.

STUDY QUESTIONS

1. Describe the components of the flow cytometer that must be tested every day to ensure consistent flow cytometric results.
2. Describe three characteristics of a good instrument standard.
3. Describe two characteristics of each of the following particles: standards, calibrators, and controls.
4. Explain how frequently optical alignment should be checked.
5. Describe the effect of partial plugs or bubbles on the cell stream and the resulting signal output.
6. Describe the effect of dirty flow cells on the data.
7. Name the components of the optical paths that converge in a properly aligned flow cytometer.
8. Describe the two basic approaches to electronic standardization of the flow cytometer.

9. Describe how the instrument sensitivity is tested using two standard particles stained with differing fluorescent intensity.
10. Describe why the instrument is adjusted to ensure that unstained cells occur in the lower left corner of a two-parameter fluorescence display.
11. Explain when secondary standards would be used in place of a primary standard for instrument setup.
12. Explain why keeping records of standards performance is important.
13. Name two optical components that when changed can affect instrument sensitivity.
14. Name three standard particles that may be used to adjust the spectral compensation network.
15. Explain why the CD8 antibody is a good choice for staining cells that are used to optimize spectral compensation.
16. Describe the potential problem that might result if a dim particle (cell) was used to adjust the spectral compensation.
17. Draw a two-parameter fluorescence display showing the location of the stained cells in each quadrant when the instrument is undercompensated, overcompensated, and correctly compensated.
18. Give a reason why a new lot of FITC-labeled antibody reagent might require a different amount of spectral compensation.
19. Describe how compensation should be checked while running the test samples.
20. Describe what should be done to qualify a new lot of reagent antibody purchased for immediate use.

IV

Quantitative Immunophenotyping in HIV Infection

INTRODUCTION

CD4+T-lymphocyte quantitation has become an important test in the staging and monitoring of persons infected with the human immunodeficiency virus. This laboratory quantitation is used to follow the progression of HIV disease and to qualify HIV-seropositive individuals for therapeutic intervention and prophylactic therapy.

Human Immunodeficiency Virus Pathogenesis

Human immunodeficiency virus infects humans by contact with infected blood, either by transfusion, intravenous drug use, or sexual contact with an infected individual. The viral particle first binds to a receptive cell surface via specific CD4 glycoproteins, then fuses to the membrane, and finally enters the cell (Fig. 4.1). The viral single-stranded (ss)-RNA is converted to double-stranded (ds)-DNA by a specific reverse transcriptase and then integrates into the nuclear DNA via specific enzymes. This incorporation into the cellular DNA causes a life-long persistent infection.

In the course of HIV infection, the metabolism of the cell decreases to a maintenance level as energy is redirected toward viral production. Viral DNA is converted back to viral RNA by the infected cell, and some viral RNA is also processed to produce the viral proteins necessary for assembly of the particles. Viral particles

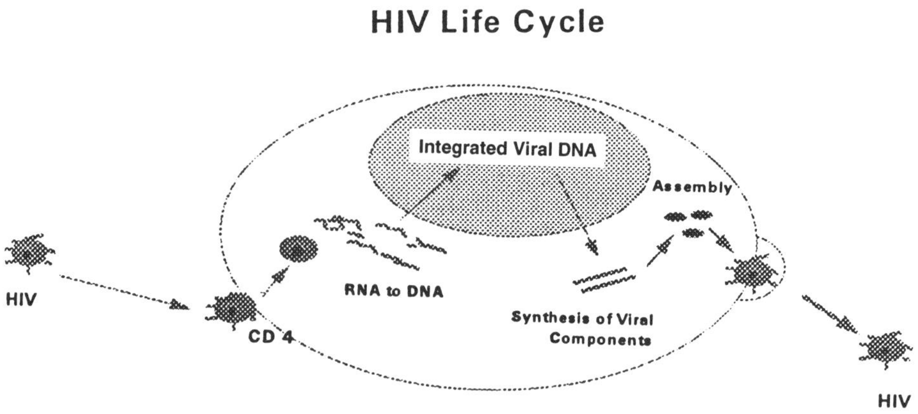

Fig. 4.1. Infection and propagation by human immunodeficiency virus. HIV binds to CD4 receptors in the cell membrane, is internalized into the cell, and takes control of the cell metabolism to produce new viral particles, which are released by budding or cell lysis.

are released from cells by budding or by lysis. Replication of the virus may result in cell death. Human immunodeficiency virus infection and intracellular replication cause the progressive loss of CD4+ T lymphocytes, with subsequent damage to immune function. A weakened immune system allows opportunistic infections to cause life-threatening diseases.

HIV host cells. All cells that are primary targets for infection by HIV carry CD4 determinants on their surface. The cells are predominantly hematopoietic, but cells bearing CD4 receptors can be found in other organ systems.

In the hematopoietic system, the T lymphocytes, specifically CD4+ T lymphocytes, are known to be a primary target for viral infection and the primary site of viral replication. Monocytes and macrophages are also known to be infected; and B lymphocytes and promyelocytes have also been suggested as possible sites of infection.

Cells from other tissues have also been found to support HIV infection. The cells that have been studied include those of the brain, skin, gastrointestinal tract, and nerves. Many of these studies involved *in vitro* cell lines, not *in vivo* isolations. Nevertheless, CD4 glycoprotein has been found in these sites during HIV disease progression. Its presence may indicate infection of these cells or the migration of CD4-bearing hematopoietic cells.

Monitoring HIV Progression

Laboratory analyses that monitor the progression of HIV disease are circulating HIV antigen quantitation, HIV-specific serology, and enumeration of lymphocyte subsets by immunophenotyping.

HIV antigen. HIV antigen, especially the p24 viral protein, is detectable prior to the production of antibodies to HIV (Fig. 4.2). The serum levels of p24 fall as HIV-specific antibody titers rise. A subsequent increase in viral p24 concentration in serum following this initial drop is indicative of a poor prognosis.

Serologic response. The immune response to HIV antigens is similar to other viral antigens. Viral-specific IgM rises following infection, then disappears as IgG titers rise. HIV-specific IgG has been reported to be detectable as early as 11 days by immunofluorescence assays. Enzyme immunoassays (EIA) detect HIV-specific IgG 30 to 60 days after the onset of acute illness and are used as a screening test. Western blot is the recommended confirmatory testing method for positive serology screens. Antibody titers rise and remain elevated until very late in the disease progression.

Cellular quantitation. Fluctuations in lymphocyte numbers occur after HIV seroconversion. CD3+CD8+ lymphocytes first increase in number, then decrease as the disease progresses. CD3+CD4+ lymphocytes experience significant decreases in number during the course of the disease. Generally, when CD4+ T-lymphocyte levels

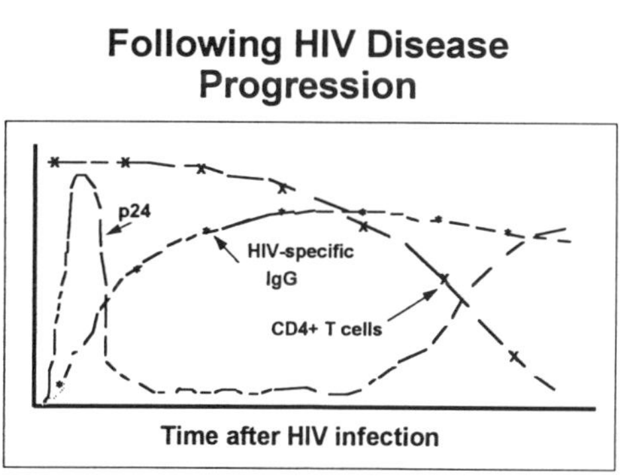

Fig. 4.2. Laboratory tests that monitor the progression of HIV disease. After HIV infection, circulating p24 antigen is the first analyte to be measured. Viral-specific immunoglobulin increases, then plateaus as the antigen becomes undetectable. CD4+ T cells decrease steadily from baseline levels prior to infection.

fall below 400–500 cells/μL of blood, there is a marked increase in the likelihood of acquiring one or more of the opportunistic diseases that define AIDS. Quantitation of CD4+ T lymphocytes is required for monitoring the progression of a person from HIV infection to AIDS.

CD4+T lymphocyte monitoring. When a patient's initial workup reveals a positive HIV serology, or signs and symptoms consistent with immunosuppression, blood should be drawn for a baseline CD4+ T-lymphocyte analysis. HIV-seropositive individuals should be followed clinically and retested every 6 months, or more often if their clinical signs change. CD4+ T-lymphocyte testing is used to determine when therapeutic antiviral intervention is warranted and is also used to monitor effects of therapy.

Classification of HIV Disease

The 1993 CDC Revised HIV Infection and AIDS Surveillance Classification for Adults and Adolescents is based on both clinical and laboratory criteria. Clinical categories reflect changes in signs and symptoms of disease and have been updated from both the 1987 Classification System and the World Health Organization (WHO) interim staging system. Laboratory categories reflect stages of immunosuppression, measured by CD4+ T-lymphocyte quantitation. Both categories are used together in staging patients, defining acquired immunodeficiency syndrome (AIDS) epidemiologically, and monitoring effects of treatment.

The revised classification allows an expanded AIDS surveillance case definition, reflects the current standard of medical practice, and provides more accurate epidemiologic data. All patients in clinical category C and all patients in laboratory category 3 are now defined as having AIDS (Fig. 4.3).

Clinical categories. There are three clinical categories of HIV disease: A, B, and C. All categories require HIV infection. The severity of disease increases from A to B, and from B to C.

Clinical category A. The first clinical category of HIV disease includes the following conditions in an adolescent or adult with *documented HIV infection:* asymptomatic HIV infection, persistent generalized lymphadenopathy (PGL, formerly called LAD), and/or any acute HIV infection with accompanying illness or history of acute HIV infection. Conditions associated with clinical categories B and C must not have occurred.

1993 Revised CDC Classification System

for HIV-Infection and Expanded AIDS Surveillance Case Definition

CD4 Categories	Clinical Categories A	B	C
(1) > 500 cells /µL (> 29% lymphocytes)	A1	B1	C1 ✱
(2) 200 - 499 cells /µL (14 - 28% lymphocytes)	A2	B2	C2 ✱
(3) < 200 cells /µL (< 14% lymphocytes)	A3 ✱	B3 ✱	C3 ✱

✱ AIDS Indicator

Fig. 4.3. Revised classification for HIV-infection and expanded AIDS surveillance case definition in adolescents and adults. In 1993, the CDC expanded the clinical definition of AIDS to include laboratory staging as well. HIV-seropositive individuals with CD4+ T-cell counts less than 200 cells/μL are now considered to have AIDS, regardless of their clinical presentation.

Clinical category B. The second clinical category of HIV disease is characterized by symptomatic complications of HIV and must meet at least one of the following criteria: (1) conditions attributed to HIV infection and/or indicative of a defect in cell-mediated immunity or (2) conditions of a disease complicated by HIV infection. Conditions associated with clinical category C must not have occurred.

Symptomatic complications may include: 1. Bacterial angiomatosis, 2. Candidiasis (vulvovaginal), persistent or poorly responsive, 3. Candidiasis, oropharyngeal (Thrush), 4. Cervical dysplasia, (moderate or severe)/cervical carcinoma in situ, 5. Constitutional symptoms (fever or diarrhea lasting more than 1 month), 6. Hairy leukoplakia, oral, 7. *Herpes zoster,* greater than two episodes or more than one dermatome, 8. Idiopathic thrombocytopenia purpura, 9. Listeriosis, 10. Pelvic inflammatory disease, particularly if complicated by ovarian tubo-ovarian abscess, 11. Peripheral neuropathy.

Clinical category C. The third clinical category of HIV disease includes conditions strongly associated with severe immunodeficiency, which occur frequently in HIV-infected individuals, and which cause serious morbidity or mortality.

Category C conditions include those listed in the CDC 1987 surveillance case definition of AIDS with several recent additions: 1. Candidiasis, bronchi, trachea, or lungs, 2. Candidiasis, esophageal, 3. Coccidioidomycosis, disseminated or extrapulmonary,

4. Cryptococcosis, extrapulmonary, 5. Cryptosporidiosis, chronic intestinal, 6. Cytomegalovirus disease, other than liver, spleen, or lymph node, 7. Cytomegalovirus retinitis, 8. HIV-related encephalopathy, 9. *Herpes simplex,* chronic ulcers, bronchitis, pneumonitis, or esophagitis, 10. Histoplasmosis, disseminated or extrapulmonary, 11. Isosporiasis, chronic intestinal, 12. Kaposi's sarcoma, 13. Lymphoma, Burkitt's, 14. Lymphoma, immunoblastic, 15. Lymphoma, primary in brain, 16. *Mycobacterium avium* complex or *M. kansasii,* disseminated or extrapulmonary, 17. *Mycobacterium tuberculosis,* any site, pulmonary or extrapulmonary, 18. Mycobacterium, any type, disseminated or extrapulmonary, 19. *Pneumocystis carinii* pneumonia, 20. Progressive multifocal leukoencephalopathy, 21. Salmonella septicemia, recurrent, 22. Toxoplasmosis of brain, 23. HIV wasting syndrome, 24. Recurrent pneumonia, and 25. Invasive cervical cancer.

Laboratory categories. Staging of the immunosuppression of HIV disease is determined by quantitation of CD4+ T lymphocytes. Patient management with CD4+ T-cell quantitation begins with an initial workup to establish the baseline level of CD3+CD4+ lymphocytes. Patients are stratified into different laboratory categories by CD4+ T-cell testing. These results are also the basis for treatment intervention and monitoring. HIV-infected individuals are stratified into the following three levels by CD4+ T-lymphocyte quantitation (or by lymphocyte percentage):

Category 1—CD3+CD4+ lymphocytes ≥ **500 cells/mL**
(>29% of lymphocytes)

Category 2—CD3+CD4+ lymphocytes **between 200 and 499 cells/mL**
(14–28% of lymphocytes)

Category 3—CD3+CD4+ lymphocytes < **200 cells/mL**
(<14% of lymphocytes)

Although treatment decisions are generally made from absolute CD4+ T-cell counts, several groups have recommended that patients be followed with CD4+ T-cell percentages (of lymphocytes) instead. These results are not subject to the variabilityof hematology measurement. While the CDC does not recommend using CD4+ T-cell precentages alone, the recommended FCI guidelines include percentage ranges as well as absolute numbers in the definition of AIDS.

T-Lymphocyte Subsets in HIV Disease

The T-lymphocyte subsets change rather dramatically during the evolution of HIV disease. The clinical staging of HIV disease progression closely follows the changes in the CD4+ T-lymphocyte values. Individuals with negative HIV serology and those with positive HIV serology but who are asymptomatic, have relatively high proportions of CD4+ T lymphocytes. Those persons who fall into clinical category A may have a decrease in the overall proportion of CD4+ T cells; patients in clinical category B may have relatively fewer CD4+ T cells; persons in clinical category C often have very low levels of CD4+ T lymphocytes (Fig. 4.4).

CD4+ T lymphocytes are the targets for the HIV virus. The virus itself identifies CD4 as a receptor and enters those cells. The progressive loss of CD4+ T lymphocytes correlates with the immunosuppression of the individual. For any individual, time course of the disease and severity of the symptoms are variable. The time course is affected by antiviral drugs such as zidovudine (AZT). The disease course may be related to the overall immune status and activation of the immune system and, therefore, may be closely related to the total number of CD4+ T lymphocytes available to protect the individual.

Interestingly, there is a compensatory increase in the CD8+ T lymphocytes as the CD4+ T lymphocytes decrease. During the early stages of the disease the total number of T lymphocytes remains relatively constant; however, the relative proportions of the T-lymphocyte subsets change rather dramatically.

Total T, B, and NK cells. The total counts of T, B, and NK lymphocytes do not correlate well with HIV disease progression (Fig. 4.5). The total T-lymphocyte count remains relatively constant throughout the course of the disease. The decrease in CD4+ T lymphocytes is offset by a compensating increase in the CD8+ T lymphocytes. This increase in CD8+ T cells is quite variable from patient to patient. Therefore,the total T-lymphocyte count is not a good measure of the disease progression.

The proportion of B cells and NK lymphocytes does not change during the course of the disease. There is an increase in total IgG

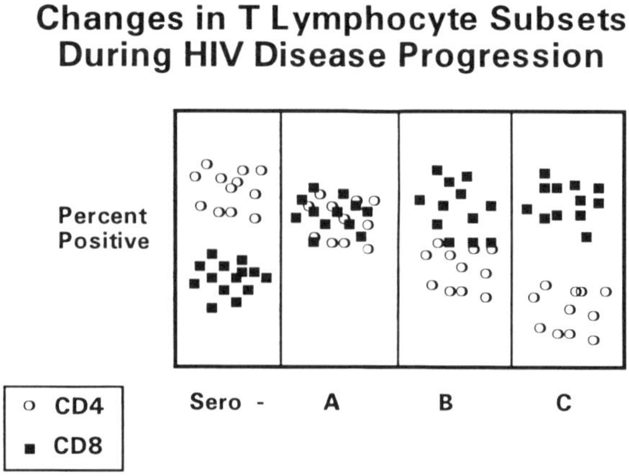

Fig. 4.4. CD4+ T lymphocytes and CD8+ T lymphocytes during HIV disease. Compared to HIV-seronegative individuals, the percentage of CD4+ T cells in HIV-infected persons begins to decrease during clinical stage A and continues to fall during stages B and C. The percentage of CD8+ T cells, in contrast, increases after HIV infection, and then plateaus.

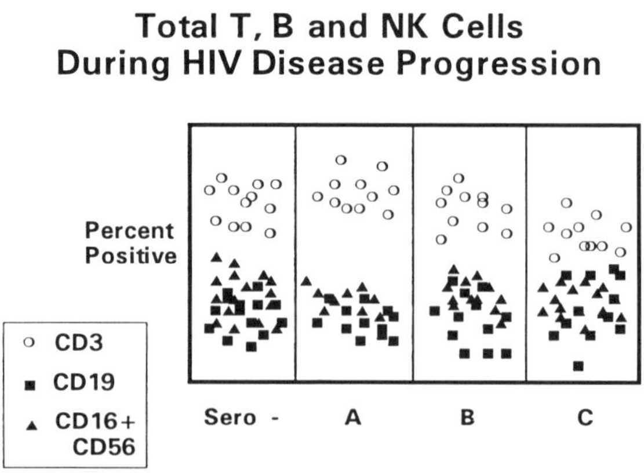

Fig. 4.5. T lymphocytes, B lymphocytes, and NK cells during HIV disease. Compared to HIV-seronegative individuals, the percentage of T cells (CD3) stays relatively constant before falling in late stages of the disease. Percentages of B cells (CD19) and natural killer cells (CD16+CD56) remain fairly constant throughout the disease.

production during HIV disease progression, but the level of serum immunoglobulin does not correlate well with the severity of the disease, although it is another indicator of progression.

CD4 : CD8 T-cell ratios. CD4 : CD8 ratios have been used to monitor HIV disease progression. The decrease in CD4 and the increase in CD8 cause a rather dramatic change in the ratio during HIV disease.

CD4 : CD8 cell ratios differ when compared with the CD4 : CD8 T-lymphocyte ratios. Determination of actual CD4+ T cells and CD8+ T cells is complicated by the presence of CD4+ monocytes and CD8+ NK cells in the lymphocyte analysis gate. The determination of CD4 : CD8 T-cell ratios is more sensitive in assessing the decline of immune function during HIV disease. The non-T-cell ratios tend to be lower than CD4 : CD8 T-cell ratios.

The CD4 : CD8 T-lymphocyte ratios can also change with other diseases, such as viral infections with cytomegalovirus (CMV) or Epstein-Barr virus (EBV). Drugs or immunosuppression may also affect the CD4 : CD8 T-cell ratio. *It is important to use serologic evidence of HIV infection along with clinical symptoms to establish that an individual is infected with HIV.* For monitoring disease progression, the CD4+ T-lymphocyte absolute counts are recommended, instead of the CD4 : CD8 T-lymphocyte ratio.

RECOMMENDED IMMUNOPHENOTYPING PANEL

The recommended panel of monoclonal antibodies for monitoring HIV disease progression consists of six two-color reagent pairs described below:

The recommended pairs of reagents provide the most accurate and unambiguous definition of the lymphocyte subpopulations. The reagents were selected to maximize information about the entire lymphocyte population. The chromophores FITC and PE were selected to give the maximum separation between populations of cells, thereby forming discrete clusters of cells. The reagents listed in Table IV.1 provide an internal quality control check for each specimen and between sample tubes.

The first two reagent combinations (CD45,CD14 and isotype control [IgG,IgG]) provide for quality control in setting the light scattering gate and in defining the negative population. The use of

TABLE IV.1 Panel of Monoclonal Antibodies for Defining Lymphocyte Subpopulations

TUBE #	FITC-AB	PE-AB
1	CD45	CD14
2 (isotype control)	IgG	IgG
3	CD3	CD19
4	CD3	CD4
5	CD3	CD8
6	CD3	CD16/56

FITC-AB: fluorescein isothiocyanate–labeled antibody; PE-AB: phycoerythrin-labeled antibody.

CD45,CD14 permits the optimal setting of the light scatter gate and quality control of that gate. It is used to maximize the number of lymphocytes included in the gate. It can also be used to determine the number and identity of nonlymphocyte contaminants in the light scatter gate. The isotype controls are used to define the position of a negative population.

The four remaining reagent combinations all use CD3 FITC in combination with other lymphocyte subset reagents. The replicate CD3 determinations can be used to identify the variability between samples from the same specimen.

Population Assignment

The goal of flow cytometry is to be able to assign any particular cell to a specific population based on its light scattering and antigenic characteristics. By assigning cells to one population or another, the composition of the blood specimen can be determined. This overall goal must be kept in mind when developing quality control procedures and rules for data analysis in lymphocyte subset and progenitor cell analyses.

Discrete populations. Isolated populations are easier to enumerate. Population assignment is facilitated when the groups of cells are completely isolated from each other and form discrete clusters (Fig. 4.6). Populations that exhibit clear differences in cellular characteristics can be readily separated from one another. The assignment of any particular cell to its population becomes trivial and the enumeration of one population relative to the entire cellular makeup

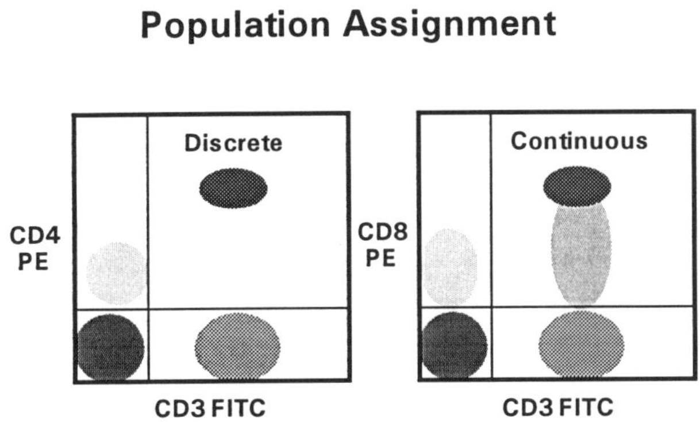

Fig. 4.6. Two-parameter histograms of discrete and continuous populations. On the left, the helper/inducer T cells (CD3+CD4+) form a discrete pattern, clearly separated from other populations. On the right, the cytotoxic/suppressor T cells form a continuous pattern, not well resolved from the negative population. For continuous populations, especially, it is important to have set the threshold between positive and negative appropriately.

is easy. The recommended panel for monitoring HIV infection was selected to maximize the distance between populations, thereby allowing discrete populations to be identified; however, the panel does have some antibody combinations that have continuous patterns.

Continuous staining. Sometimes populations of positive-staining cells are not well resolved from the negative-staining cells (Fig. 4.6). This situation occurs when antigen expression is dim and not well resolved from the population of cells that do not express the antigen. Within the recommended HIV reagent panel, there are problems with identifying dim-staining CD8+ cells and CD56+ T cells. Continuous-staining populations are often seen with activation antigens such as HLA-DR, CD25, and CD38. Although these reagents are not part of the recommended panel for monitoring HIV infection, they are often requested by physicians in order to determine the number of activated T lymphocytes within a specimen.

More care is needed in differentiating positive cells from the negative population in continuous-staining populations. Selection of the proper negative control becomes important and it may not be appropriate to simply use isotype-matched reagents. For example, for the enumeration of CD3+CD25+ cells it would be better to use a negative reference population from the CD3,CD19 combination of reagents. In this way the CD3+ cells alone will act as a reference for determining the CD3+CD25+ cells.

Instrument sensitivity and compensation. Discrimination between dim and negative cell populations can be affected by low instrument sensitivity and by improper spectral compensation. Poor instrument sensitivity or improper spectral compensation settings may result in underestimation of continuous-staining populations. Therefore, more care must be taken to ensure adequate instrument performance when assaying samples with continuous rather than discrete populations of cells.

Examples of discrete and continuous staining. An example of a reagent combination that has discrete populations is CD3,CD4. The CD3+ cells are easily identifiable from those cells that do not express this antigen. The population of CD3+ cells is split into two non-overlapping groups based on CD4 expression (CD3+CD4+ and CD3+CD4−). It should be noted that in some samples there are some cells that express CD4 dimly; however, these are CD3− monocytes. With this combination of reagents all populations are easily separated.

An example of a continuous population can be found with CD3,CD8. Again the CD3+ cells are easily separated from the cell population that does not express CD3. The CD3+CD8+ cells, however, are continous with the CD3+CD8− population. Accurate discrimination is more difficult in this case and requires careful use of the negative reference population and proper spectral compensation. In this case the best negative reference population would be identified using the CD3,CD19 reagent combination.

Continous populations can also be observed in specimens that have been stored or shipped improperly. They are also observed in specimens that are older and have increased nonspecific staining. In such cases, it is essential to use the proper negative population to set the markers.

Isotype Controls

The isotype control reagents are used to define the position of the negative cells and to set the fluorescence markers (Fig. 4.7). The negative population must be carefully defined in order to distinguish between positive and negative populations. The percentage of positive cells should be limited to $<2\%$ in the single-positive and double-

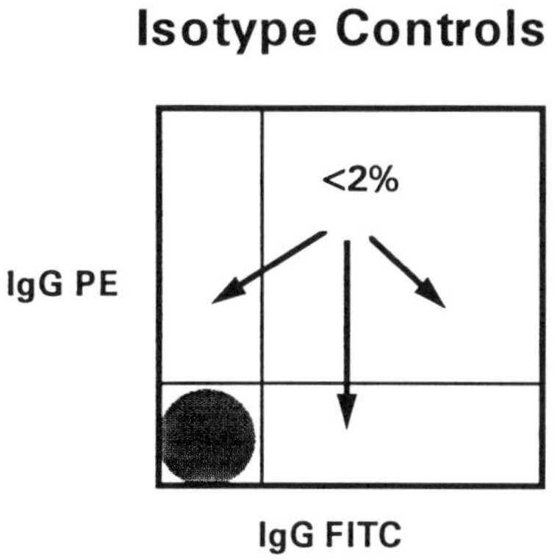

Fig. 4.7. Setting fluorescence markers on a negative population using isotype controls. Using two monoclonal antibodies without reactivity to the test population of cells, fluorescence thresholds can be set to contain all negative cells. The isotype control contains both nonspecific binding and autofluorescence.

positive quadrants. Two antibody controls that do not bind to human cells are used to determine the amount of nonspecific staining of these antibodies with the target populations. This approach is primarily a holdover from the days of single-color flow cytometry and the use of polyclonal antibodies.

Nonspecific staining. An additional step that permits a measurement of nonspecific staining is to run a tube containing unstained cells in parallel with the isotype control. The position of the unstained cells (autofluorescence) can be compared to the cells prepared with the isotype control. The separation between the populations (observed by overlaying the two plots) gives a measure of the relative nonspecific binding of irrelevant antibodies to the cell populations.

Isotype control equivalent to test reagents. The use of isotype controls as a reference population assumes that the same nonspecific staining and biochemical properties are present as in the test reagents. There are differences between immunoglobulin isotypes in their binding to Fc receptors. Different types of Fc receptors are found on different cells. These receptors have different affinities for the various heavy chains of the immunoglobulin molecule. It is known that dead cells have much more nonspecific staining than do intact cells. It should be noted that the use of isotype controls may not give an exact match for each of the reagents in the panel. There may be differences in isotype (heavy chain), color, or concentration between the control and the test reagents.

Differences in antibodies. In addition it should be noted that conjugation of FITC to an antibody changes the overall charge of the complex. Overconjugation may significantly change the binding characteristics of the antibody. Differences in the amount of aggregation among antibodies may cause variability in nonspecific staining. Therefore, the position of the negative cells (appearing in the lower left-hand corner of the plot) should be noted and used as an additional internal quality control check for each sample tube. The negative cells in tubes 3 through 6 (Table IV.1) should appear in the same position as the negative population for the isotype control.
Positive-staining reference. Isotype controls are less important when the reagent panel is comprised of reagents yielding discrete populations. Additional controls are required to carefully distinguish between single-positive and double-positive cells. The use of an isotype (double-negative) control may not be appropriate, especially for reagents that identify groups of cells that stain with two colors or more. A more appropriate control would be reagents that define accurately the green-only and the orange-only populations of cells. Therefore, an appropriate reference population may be better identified using CD3,CD19 rather than an isotype control reagent.

T-Lymphocyte Identification

Mature T lymphocytes are identified by the expression of CD3 on their cell surfaces. Although several markers have been used historically to identify T lymphocytes, CD3 has been shown to be the most specific antigen. This glycoprotein is one of five proteins in the T-cell receptor complex and is only expressed on T cells.

CD3 is expressed on the cell surface relatively late during the development of the T lymphocytes in the thymus. Therefore, it can only be used to identify mature T lymphocytes. Selection of this reagent for monitoring HIV infection was based on the finding that the mature T lymphocytes are the ones that are lost during progression of HIV infection.
CD3 FITC. The FITC conjugate of CD3 provides a discrete separation of the T lymphocytes from the negative cells in essentially all specimens. Therefore, CD3 FITC was selected as the lineage identifier in the panel of reagents monitoring HIV infection. It is not necessary to have CD3 conjugated to PE to give added separation between the negative and the positive populations.

Replicate CD3 testing. In the recommended panel of reagents, CD3 FITC is used in four different antibody combinations to stain samples. Replicate use of CD3 permits identification of discrete populations and allows an internal assessment of the between-tube variation. The reagents labeled with PE stain more brightly than those labeled with FITC. Therefore, PE-labeled antibodies have a greater separation between positives and negatives than FITC-labeled antibodies. The antibody combinations were chosen to enhance the separation between the lymphocyte subpopulations being analyzed.

T (CD3) and B (CD19) cells. T and B lymphocytes are identified using CD3,CD19 (Fig. 4.8). CD3 is the most specific reagent to identify T lymphocytes. With this combination of reagents, no cells should be positive for both CD3 and CD19. Therefore, the CD3+ cells are green only and provide an ideal reference population to compare with the CD3+ cells in the tubes stained with the remaining reagent combinations in the panel.

CD19 is the most specific reagent to identify B lymphocytes. It is expressed very early during the development of B lymphocytes and is lost just prior to the formation of plasma cells after antigenic stimulation.

CD3 and CD19 define discrete separated populations. Any double-positive cells are artifacts of the sample or specimen. Often

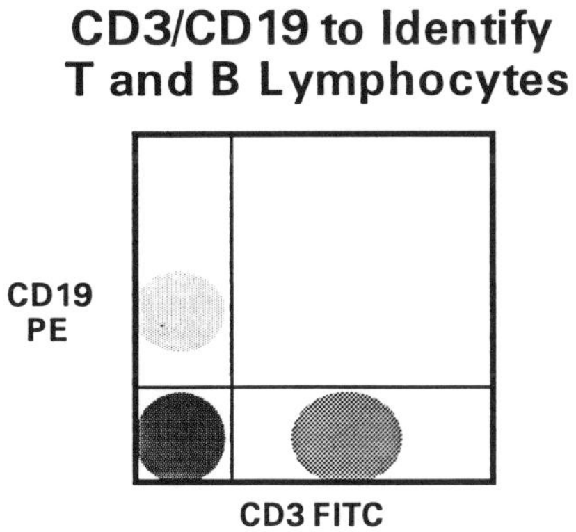

Fig. 4.8. Two-parameter fluorescence histogram to identify T and B lymphocytes. CD3+ cells form a discrete population of T cells, easily discriminated from the negative population. CD19+ cells form another discrete population, B cells, also well-resolved from negatives. Any double-positive cells would appear in the upper right quadrant. Negatives, in the lower-left quadrant, contain NK cells and any nonlymphocyte contaminants in the gate.

a few double-positive cells can be identified in which a T cell and B cell form a doublet. In this way, two lymphocytes may stick together and flow through the instrument as if they were one large cell.

Quality control check. The combination of CD3, CD19 can be used as a quality control check for compensation, as well as for nonspecific staining. This reagent combination should display three discrete populations at right angles to each other. If the populations are not discrete or if the populations are curved (not orthogonal), a flag should be noted on the sample and/or specimen for further, more careful analysis. This is also important for the quality control of all subsequent reagent combinations in this panel. It must be assumed that if there is a problem with instrument performance for one tube, then all the other tubes in the panel may be affected. For those tubes that have continuous staining populations, it is difficult to observe this type of problem. Therefore, the CD3,CD19 tube can be used as an additional quality control check for the entire panel of tubes being analyzed.

CD4+ T lymphocytes (CD3,CD4). The CD4+T lymphocytes are identified using CD3,CD4 (Fig. 4.9). The T cells are identified by staining with CD3. The double-positive cells, CD3+CD4+, are the true CD4+ T lymphocytes (helper/inducer). These are the primary target of HIV infection. It is this population that decreases during the disease progression and immunosuppression observed following HIV infection. Clinically this is the most important cell population to monitor.

CD4+ monocytes. The CD3−CD4+ population is comprised of monocytes. These cells express CD4 dimly as compared with the CD4+ T lymphocytes and may be more autofluorescent than lymphocytes. These cells are included in the analysis to varying degrees, based on the purity of the light scatter gate. The combination of CD3,CD4 allows for the complete separation and identification of lymphocytes from the CD4+ monocytes. Because of this complete separation, it is better to have a large light scatter gate that includes all of the lymphocytes, rather than a tight, more pure lymphocyte light scatter gate. Inclusion of some monocytes into the light scatter gate will not affect the data, since monocytes can be easily distinguished from the CD3+CD4+ T lymphocytes. Platelets and red cell debris are identified within the double-negative CD3−CD4− population.

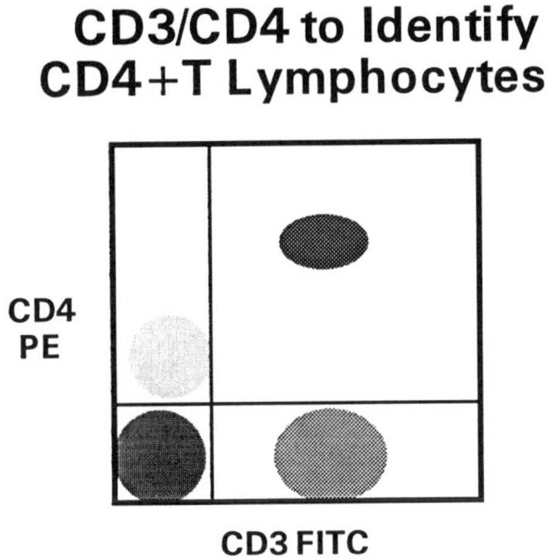

Fig. 4.9. Two-parameter fluorescence histogram to identify CD4+ T lymphocytes. Cells that are CD3+ only again form a discrete population, easily discriminated from negatives. CD3+ cells that are also CD4+ form another discrete population, in the upper right quadrant. CD3− cells include the CD4+ monocytes (upper left quadrant) and the B cells and NK cells (lower left quadrant).

Quality control check. The CD3,CD4 reagent combination can also be used as another quality control check for compensation and nonspecific staining. The double-negative population should overlap exactly with the negative population in the isotype control. The CD3+CD4− population should also overlap identically with the CD3+ population in the CD3,CD19 reagent tube.

CD8+ T lymphocytes (CD3,CD8). The CD8+ T lymphocytes (cytotoxic/suppressor) are identified using CD3,CD8 (Fig. 4.10). The population of cells that expresses both markers, CD3+CD8+, are the true CD8+ T lymphocytes and include both dim and bright CD8+ populations. The CD8 distribution is continuous with that of the negative cells. Therefore, the value obtained for the proportion of CD8+ T cells depends upon the instrument sensitivity and the establishment of correct compensation.

CD8+ NK lymphocytes. The CD3−CD8+ cells contain a subset of NK cells and a few granulocytes. These cells tend to be dim CD8+ and are easily distinguished from the CD3+CD8+ T lymphocytes.

Quality control check. This combination of reagents (CD3,CD8) can be used as an additional quality control check of compensation

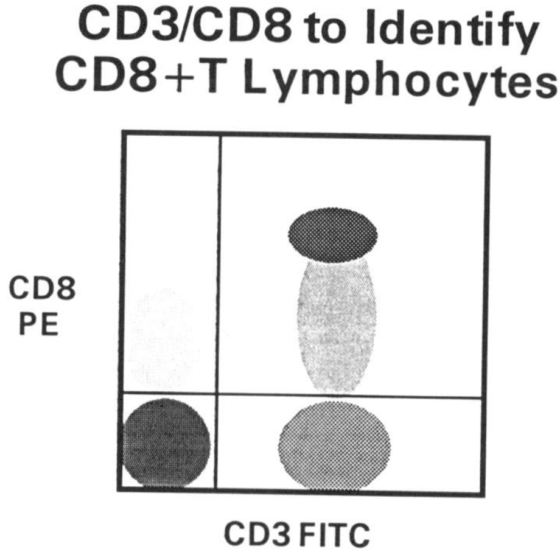

Fig. 4.10. Two-parameter fluorescence histogram to identify CD8+ T lymphocytes. Cells that are CD3+ only (lower right quadrant) are not well resolved, with the continuous population of CD3+CD8+ cells in the upper right quadrant. Cells that are CD8+ only represent a subset of NK cells. The negative population includes B cells and other NK cells.

and nonspecific staining. The double-negative population should overlap with the isotype control. The CD3+ population should appear in exactly the same spot as with the CD3,CD19 and CD3,CD4 tubes.

NK lymphocytes (CD3−,CD16+ and/or CD56+). Natural killer cells are more difficult to distinguish from T lymphocytes. CD16 (FcRIII) is expressed on most NK cells but is also expressed on neutrophils. This antigen is lost during activation of NK cells and is expressed relatively dimly. CD56 is expressed on most, but not all, NK cells, as well as on some T lymphocytes. When CD56 is used in combination with CD3 it is possible to distinguish between the CD3+CD56+ T lymphocytes and the CD3−CD56+ NK cells.

The best identification of all NK lymphocytes requires the use of all three monoclonal antibodies in combination. The NK cells express either CD16 or CD56 but do not express CD3. The combination of CD16 and CD56 also helps to separate the NK cells in intensity from the double-negative populations. In this way, they form a discrete cluster, distinct from the double-negative cells (Fig. 4.11).

CD3/(CD16+CD56) to Identify NK Cells and T Lymphocytes

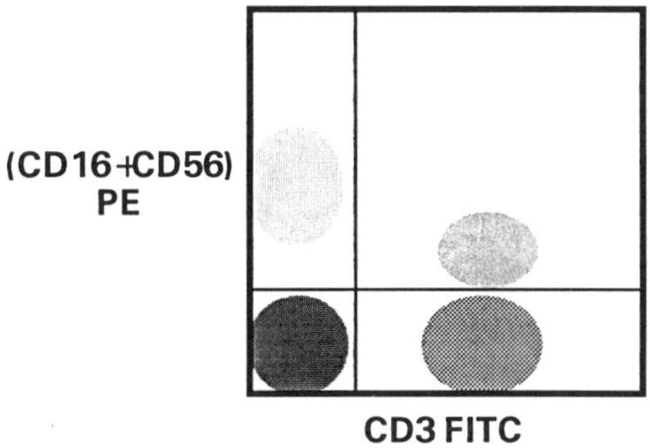

Fig. 4.11. Two-parameter fluorescence histogram to identify T lymphocytes and NK cells. Natural killer cells form a discrete cluster of CD3−, CD16+CD56+ cells in the upper left quadrant. T lymphocytes are CD3+, including both the lower right quadrant and the CD16+CD56+ T cells in the upper right quadrant. The negative cell population includes B lymphocytes and any nonlymphocytes in the light scatter gate.

GATING AND ENUMERATION OF POPULATIONS

Electronic gating is used to set analysis regions for immunophenotyping. First, the population is identified by light scattering and fluorescent-antibody reactivity. Then, subsequent monoclonal reagents are tested for binding to the same population of cells.

Lymphocyte Enumeration

Cells are determined to be lymphocytes by satisfying a four-parameter definition. Two parameters of light scatter and two parameters of fluorescence combine to identify this population for further analysis.

Determining leukocyte antigen expression. For HIV monitoring, the quantitation of lymphocyte subsets in peripheral blood is the clinically useful feature to be obtained from FCI. It is important to have an accurate enumeration of the lymphocytes in each specimen. Although ***"It is assumed that the monoclonal antibody reactivity is cell-type specific, unambiguous, and reproducible,"*** antigens identified by the monoclonal antibodies are expressed on several different cell types, not just lymphocytes. The goal is to determine the proportion of lymphocytes that express specific antigens as distinct from other cell types. For complete lymphocyte analysis, it is necessary to aliquot the specimen into several sample tubes for analysis with six different combinations of two-color monoclonal antibody reagents.

The common characteristic between samples is light scattering. Since several samples need to be analyzed for each specimen, comparisons must be made through common characteristics. The unifying feature in correlating immunophenotyping results is the light scattering pattern. As ***"It is assumed that the common characteristics between samples are maintained,"*** it is assumed that the light scattering is identical for each sample within a specimen. However, the nonlymphocytes within a specimen have different reactivities with different antibody combinations. Therefore, it is important to know how many nonlymphocytes are included in the analysis window (lymphocyte gate) and their identity. This information must be used to correct the subsequent analyses.

Hematology analysis is required to determine lymphocyte proportions. It is also important to know the proportion of lymphocytes in each specimen in order to calculate absolute counts. The absolute number of lymphocytes can be determined using an automated hematology analyzer. This absolute number of lymphocytes can then be used with the percentages of subsets obtained from flow cytometric data to calculate the absolute number of each subset of lymphocytes.

Characterizing the cells in the lymphocyte analysis gate. It is possible to use immunofluorescence to aid in setting the proper light scatter gate. By using fluorescence, it is possible to optimize the light scatter gate to include the maximum number of lymphocytes and to minimize the number of nonlymphocytes in the gate. Fluorescence can also be used to identify the nonlymphocytes within the light scatter gate.

CD45 and CD14 Expression

CD45 and CD14 antigens have different expression pattens on different cells. All leukocytes express the CD45 antigen, the common portion of the T200 antigen. Lymphocytes express high levels of

CD45, monocytes express intermediate levels of CD45, and granulocytes express low levels of CD45.

Since all leukocytes express CD45, it is not appropriate to use the negative control to set markers in the distinction of lymphocytes, monocytes, and granulocytes. CD14 is expressed in high levels on monocytes and is dimly expressed on the mature neutrophils. Lymphocytes are negative for CD14 although a small subset may express this antigen dimly.

CD45 histogram. The differential expression of CD45 on leukocytes is evident in examining the CD45 histogram (Fig. 4.12). There are significant differences in CD45 expression on lymphocytes, monocytes, and granulocytes; however, a single parameter is not sufficient to distinguish these cells because of the significant overlap in the intensity distributions. The degree of separation between cell types depends on the specific CD45 antibody clone used in the analysis.

Combining CD45 with CD14. By combining CD45 and CD14, it is possible to identify the major lineages of mature cells in peripheral blood using fluorescence alone (Fig. 4.13). The lymphocytes, monocytes and granulocytes can easily be distinguished based on their immunofluorescent properties.

Combining CD45,CD14 with light scatter. By combining CD45 and CD14 with the light scatter parameters, it is possible to identify even the minor components of mature cells in peripheral blood (Fig.4.14.). The lymphocytes express the highest levels of CD45 and little or no CD14. These cells have low forward light scatter

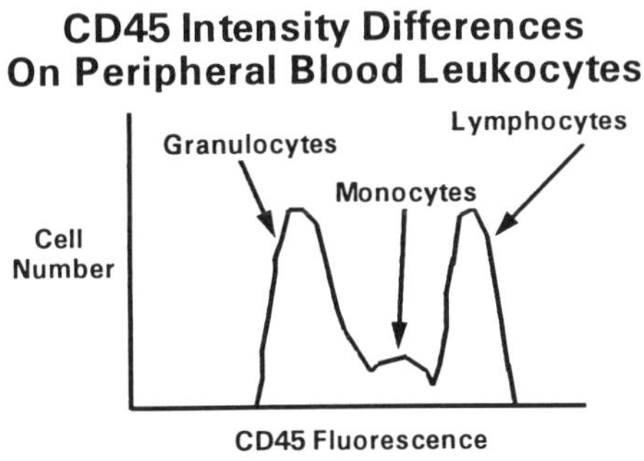

Fig. 4.12. Fluorescence intensity differences of CD45 on three populations of peripheral blood leukocytes. Lymphocytes stain the most intensely with CD45, monocytes have an intermediate intensity, and granulocytes stain the least with CD45.

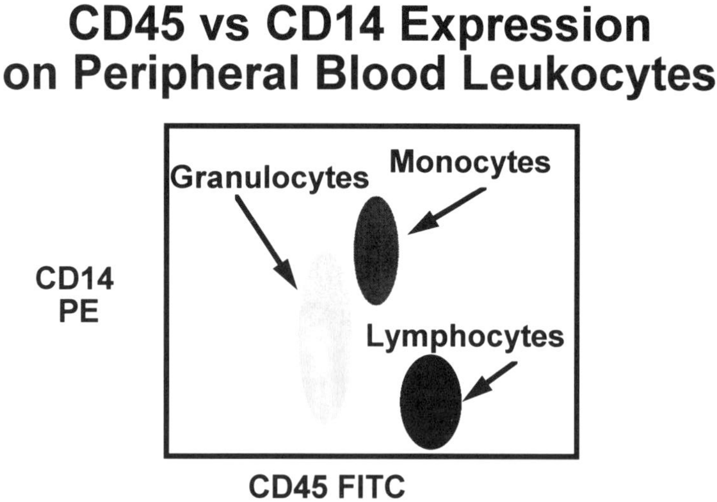

Fig. 4.13. Identifying three major populations of peripheral blood leukocytes with CD45 and CD14 immunofluorescence. CD14, in combination with CD45, separates monocytes from lymphocytes and granulocytes in a two-parameter histogram. With immunofluoresence alone, all three populations can be discriminated.

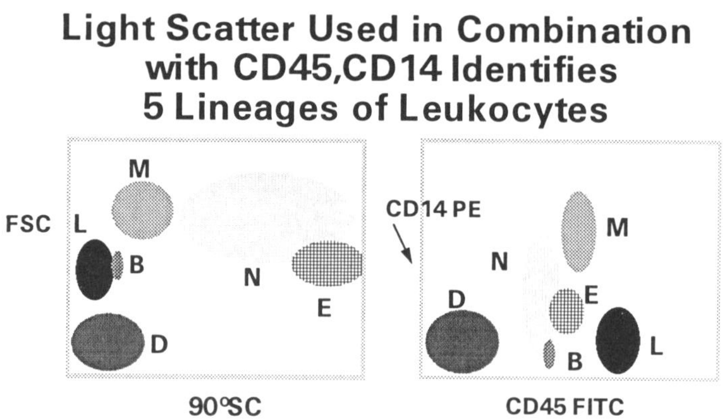

Fig. 4.14. Identifying five lineages of leukocytes with forward scatter, side scatter, and CD45 and CD14 immunofluorescence. A combination of light scatter and fluorescence can be used to identify clusters of basophils, eosinophils, and neutrophils, in addition to monocytes and lymphocytes.

and low right-angle light scatter. The monocytes express high levels of CD14 and intermediate levels of CD45. They are larger by forward light scatter and have increased right angle light scatter. The neutrophils express low levels of CD14 and CD45. These cells are large by forward light scatter and have high levels of right angle light scatter.

Eosinophils express intermediate amounts of CD45, very high right-angle light scatter and are smaller by forward light scatter than the neutrophils. The basophils have light scattering properties that are almost identical to those of lymphocytes. They express low levels of CD45 and low levels of CD14. Platelets and erythrocytes are distinguished from the leukocytes by the absence of CD45 expression.

Lymphocyte Purity and Recovery

In setting the optimal light scatter gate it is assumed that lymphocytes can be uniquely and completely identified by light scattering. It is necessary to have *high purity* with only lymphocytes included in the gate. It is also important to include *all* the lymphocytes within that gate so as not to skew the results by excluding any lymphocyte subset from the analysis. On one hand, it is necessary to have high recovery of lymphocytes and on the other hand, it is important to also have high purity. These may be conflicting goals, since a large light scatter gate is required for maximal recovery while a small light scatter gate is necessary for maximal purity.

Optimizing the light scatter gate. Several methods can be used to combine fluorescence and light scatter to optimize the gate. The method of choice depends on the software available for the instrument.

Method 1. The first step in setting the optimal light scatter gate is to identify the lymphocytes, using CD45,CD14 fluorescence (Fig. 4.15). Display a correlated two-parameter dot plot of CD45 FITC versus CD14 PE fluorescence. Set a gate around the bright CD45+CD14− cell population. Draw a large gate in order to include all of the lymphocytes in the sample.

Using this analysis gate, display a light scatter plot of the cells within this gate (Fig. 4.16). The lymphocytes will be easily distinguished from the granulocytes that might be included in the fluores-

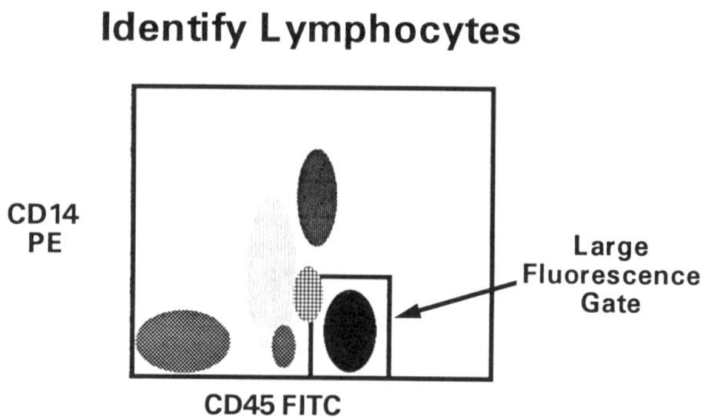

Fig. 4.15. Method 1: Setting the light scatter gate. A large gate should be drawn to include all lymphocytes, defined as CD45 bright and CD14−. This gate may include some eosinophils and neutrophils.

cence gate. Set a second gate around the lymphocytes to determine the total number of lymphocytes in the sample. These are the cells that meet both the fluorescence and light scatter criteria.

Determine the light scatter gate purity by gating on the light scatter gate and examining the CD45 versus CD14 fluorescence (Fig. 4.17). Determine the purity of lymphocytes in the light scatter gate relative to all events in that gate.

If the purity is <85%, then the light scatter gate must be reduced in size (Fig. 4.18). Determine the recovery and purity of the smaller

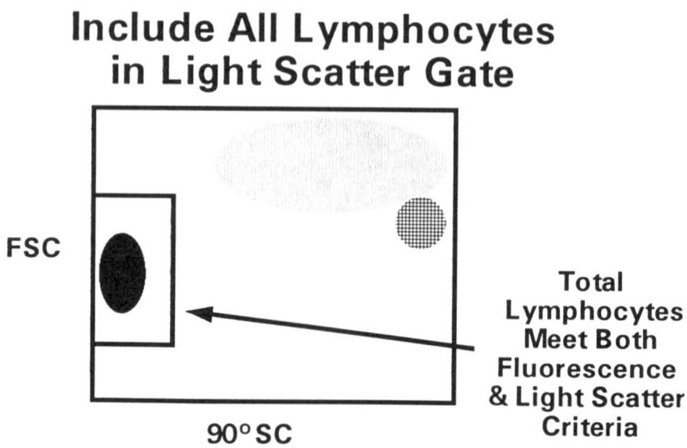

Fig. 4.16. Method 1: Determining total lymphocytes. Gating on lymphocytes from the CD45,CD14 fluorescence histogram, display the light scatter gate to separate granulocytes from lymphocytes. Set a second gate on the lymphocytes and determine the total number of lymphocytes in the sample, based on a four-parameter definition of a lymphocyte.

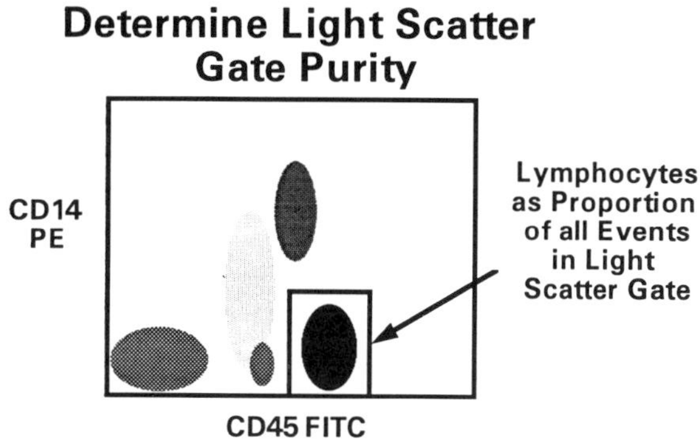

Fig. 4.17. Method 1: Determining light scatter gate purity and recovery. Draw a smaller light scatter gate for analysis. The purity of this gate is assessed by displaying the CD45,CD14 histogram of the gated events and calculating the proportion of lymphocytes in the display. The recovery is the number of lymphocytes in the gate divided by the total lymphocytes in the sample.

light scatter gate. The recovery must be ≥90% and preferably ≥95%. The purity of lymphocytes in the light scatter gate must be ≥85%. Adjust the light scatter gate to include the maximum number of lymphocytes while reducing the other contaminating cells.

Method 2. A second method uses sequentially smaller light scatter gates to maximize the inclusion of lymphocytes (Fig. 4.19). Set a large light scatter gate to include all lymphocytes. This large light scatter gate should include some debris and some monocytes, but must include all lymphocytes. Determine the number of lympho-

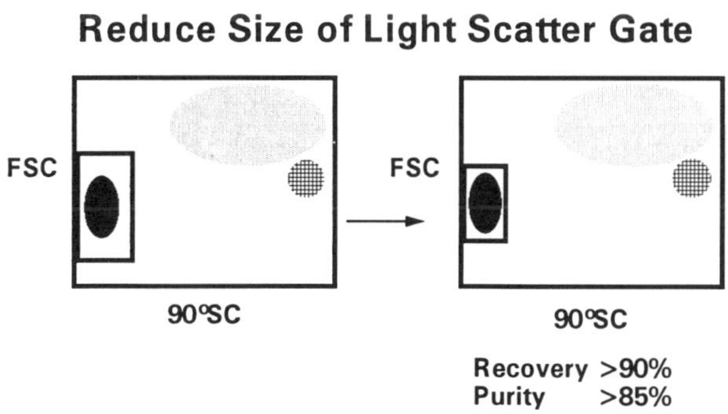

Fig. 4.18. Method 1: Improving purity by reducing the size of the light scatter gate. If lymphocyte purity of the analysis gate is <85%, a smaller gate should be drawn and the sample reanalyzed for both lymphocyte recovery and purity. Recovery should be >90% and purity >85%.

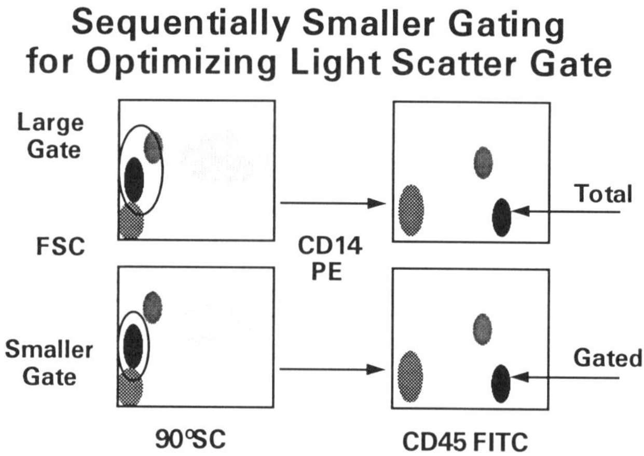

Fig. 4.19. Method 2: Sequential reduction in size of the light scatter gate. After the total lymphocytes have been enumerated from a large light scatter gate, reduce the size of the gate and calculate recovery and purity. Continue reducing the size of the gate until this analysis gate gives the best combination of recovery (>90%) and purity (>85%).

cytes that fall into this gate and express high levels of CD45, but do not express CD14.

The light scatter gate must be reduced in size if the purity is <85%. Set a smaller light scatter gate to exclude the majority of debris and monocytes. Determine the number of lymphocytes in this smaller gate. The light scatter gate should be set to include ≥90% (preferably ≥95%) of the total lymphocytes and be ≥85% pure lymphocytes.

Method 3. The first step in this method is to identify the leukocytes using CD45 in combination with forward light scatter (Fig. 4.20). Set the display to show the correlation between CD45 and forward light scatter. The cells that are negative for CD45 are debris, platelets, and erythrocytes. All of the CD45+ cells are leukocytes. The brightest CD45 cells are lymphocytes.

Set the forward light scatter threshold (discriminator) to include all lymphocytes, excluding as much debris as possible. Setting the forward light scatter trigger using this display ensures that the smallest lymphocytes are not excluded from the analysis.

The monocytes can be distinguished from the lymphocytes based on their expression of CD14 and their right angle light scatter. Set the display to show CD14 versus right angle light scatter (Fig. 4.21).

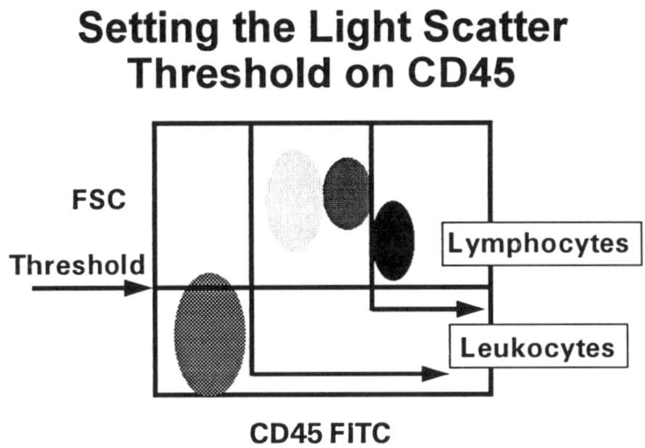

Fig. 4.20. Method 3: Setting the foward light scatter threshold. On an ungated display of forward light scatter and CD45, set the forward light scatter threshold to include all lymphocytes and to exclude as much debris as possible. All leukocytes will be included above this threshold.

The monocytes are bright for CD14 with increased right angle light scatter. Lymphocytes are negative or dim for CD14. Granulocytes have high right-angle light scatter and express CD14 dimly.

Set the right angle light scatter gate to include all lymphocytes while excluding the majority of monocytes. This light scatter gate should include the maximum number of lymphocytes. It should be

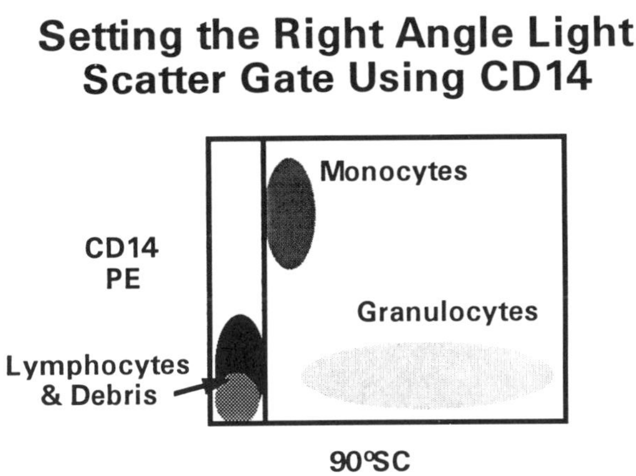

Fig. 4.21. Method 3: Setting the side scatter threshold. On the same ungated data as the previous figure, display CD14 and 90° scatter. Set the side scatter threshold to include all lymphocytes but as few monocytes as possible. Basophils and debris will also be included in the lymphocyte region.

noted that some debris (platelets and erythrocytes) and basophils will also be included with the lymphocytes.

These two gates will form a rectangle when the forward and right angle light scatter signals are displayed (Fig. 4.22). Determine the recovery and purity of lymphocytes within this gate. The recovery must be ≥90% (preferably ≥95%) and the purity must be ≥85%. If these proportions are not attained, the light scatter gate must be reduced in size.

Lymphocyte recovery. Calculate the recovery of the lymphocytes in the smaller gate from: (1) the total number of lymphocytes and (2) the number of lymphocytes included in the final (smaller) light scatter gate. The total number of lymphocytes (per 10,000 events) is determined with a large light scatter and fluorescence gate that includes all lymphocytes. The gated lymphocytes (per 10,000 total events) are determined from the smaller gate. The proportion of lymphocytes included in the smaller gate is equal to the gated lymphocytes times 100, divided by the total number of lymphocytes.

Enumerating gated nonlymphocytes. Once the light scatter gate has been optimized to include the largest proportion of lymphocytes, it is possible to identify the other cells that are included within the light scatter gate (Fig. 4.23).

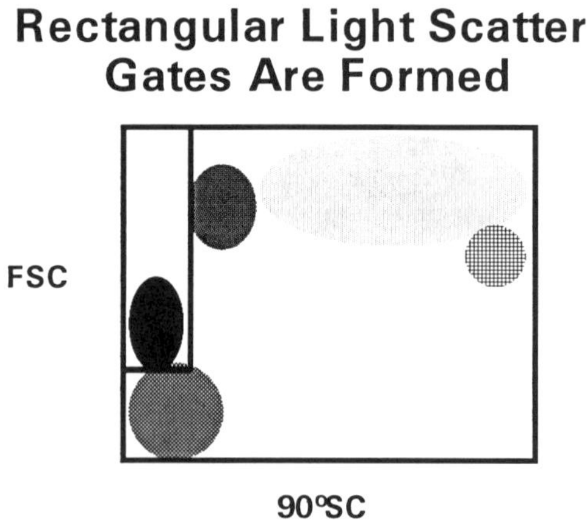

Fig. 4.22. Method 3: Completing the light scatter gate. Display forward light scatter and side scatter with the new thresholds. A rectangular light scatter analysis gate will be formed. Calculate the recovery and purity of this gate; reduce the size of the gate if the recovery is <90% or the purity <85%.

Identifying Nonlymphocytes in Light Scatter Gate

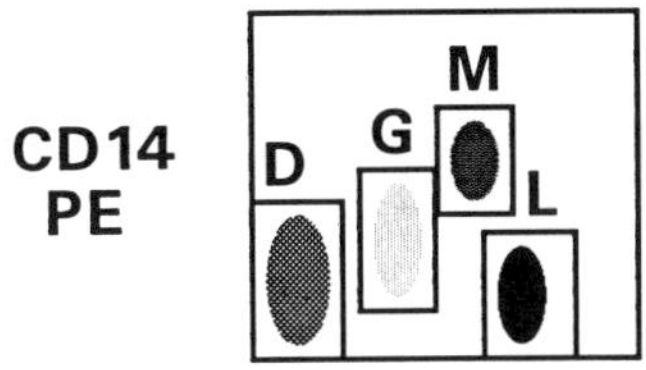

Fig. 4.23. Identifying nonlymphocytes in the light scatter gate. Gating on light scatter, display CD45 and CD14 fluorescence. Identify lymphocytes, monocytes, granulocytes, and debris clusters and draw an analysis region around each. Calculate the total number of events in the gate and the proportion of each subpopulation in the gate.

With the light scatter gate set, display the CD45,CD14 fluorescence. Determine the total number of events in the gate. Determine the number of lymphocytes in the gate that express high levels of CD45 and low levels of CD14. Divide the number of lymphocytes in the gate by the total events in the gate to determine the *purity* of the gate. Use this number to correct the subsequent subset percentages obtained. Quality control specifications require the purity to be ≥85%.

Use the CD45,CD14 fluorescence to identify the lineage of the other cells included within the lymphocyte gate. Determine the proportion of monocytes, granulocytes, and debris in the light scatter gate. Correction of subsequent immunofluorescence data may require subtracting one or more of these proportions from some of the data.

Light scatter bias. The immunophenotyping data can be biased by setting the light scatter gate too small. Lymphocytes are a heterogeneous population and not all lymphocytes scatter light identically. The CD4+ T lymphocytes and the B lymphocytes scatter less light than do the CD8+ T lymphocytes and the NK lymphocytes.

By excluding part of the lymphocytes based on light scatter, it is possible to introduce bias into the data (Fig. 4.24). A tight lymphocyte gate set to exclude debris will also eliminate some of the smaller lymphocytes, B cells, and CD4+ T cells. A tight gate set to eliminate monocytes will preferentially exclude NK cells and

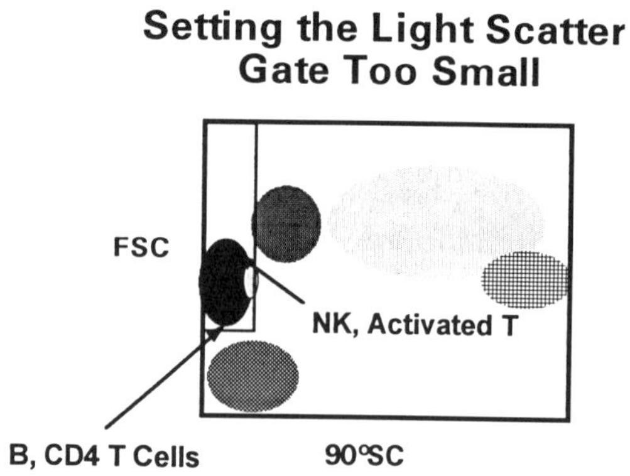

Fig. 4.24. Excluding lymphocyte subsets with a small light scatter gate. Because lymphocytes are a heterogeneous population of cells, certain subpopulations can be excluded from analysis if the light scatter gate is set to exclude either the largest (activated T and NK cells) or the smallest (B and CD4+ T cells) lymphocytes.

activated T cells. Therefore, the goal must be to set a large lymphocyte gate and to correct all results for the purity of the gate.

QUALITY CONTROL SPECIFICATIONS

The selection of the reagents for the recommended panel provides for significant internal quality control for each specimen.

Replicate CD3 Determinations

Replicate CD3 determinations assess between-tube variation. The difference between the CD3 determinations should not exceed the variance defined within the laboratory. The difference between the highest and lowest values of CD3 for any specimen should not exceed 5%. If the highest and lowest values are greater than 5%, the data should be flagged for further review.

Identifying Outlier Samples

If a specimen has been flagged, it is usually easy to identify which of the samples is suspect. In most cases, three of the four CD3 values are very close. The outlier can then be identified for more careful analysis. There are several important points to be checked.

First, is the light scatter identical for each sample tube? It is possible that the light scatter pattern for the outlier tube may be different from the others. Were the proper reagents added to the right sample tubes? What about the position of the fluorescence markers or boundaries between the negatives and positives? It is possible that on the outlier sample the boundaries are not appropriate for that particular tube.

If the results are still outside of the 5% specification, then that outlier sample must be rejected. If all other quality control criteria are met, it is possible to report the results for the remaining sample tubes.

Accounting for All Lymphocytes

The reagent panel recommended for monitoring HIV disease progression allows for the identification of all lymphocytes. The panel identifies all the T, B, and NK lymphocytes in the sample; therefore, the sum of the percentages for T, B, and NK lymphocytes should equal 100%.

T cells are defined using the average of the four CD3+ counts in tubes 3 through 6 of the panel (Table IV.1). The B cells are identified as CD19+ and CD3− in tube number 3 of the panel. The NK lymphocytes are CD3− and either CD16+ or CD56+. Optimally, the sum of the percentages for CD3+, CD19+, and CD3−, (CD16+/CD56+) should equal 100 ± 5% of the lymphocytes defined by CD45, CD14. Minimally, the variability should be ±10%.

The summation of lymphocytes uses data from all six tubes. The first tube in the panel is used for assessing nonlymphocyte contamination of the lymphocyte analysis gate and the isotype control is used to determine fluorescence marker placement. Therefore, the lymphocyte summation will only be valid if all tubes demonstrate consistency in the results. The summation requires a correction for nonlymphocytes in the light scatter gate.

Nonlymphocyte contamination. Nonlymphocyte staining should be distinguished from lymphocyte staining in the light scatter gate. It is important to correct the data for nonlymphocytes that contaminate the light scatter gate.

Monocytes are dimly positive for CD4 but do not express CD3. Monocytes also exhibit an increase in autofluorescence and/or nonspecific staining that may often be detected by a slight shift into the double-positive quadrant. It may be necessary to move the fluorescence marker in the FITC channel if a large number of monocytes are included in the light scatter gate.

Granulocytes usually cause problems in the analysis of samples that have not been stored or shipped properly. Dead or dying granulocytes have increased nonspecific staining and also have increased autofluorescence relative to lymphocytes. This will cause an increase in the double-positive cells for certain samples. Neutrophils express CD16 and may be identified as NK cells if they fall within the light scatter gate. Basophils are a small population but have the same light-scattering characteristics as lymphocytes. The basophils are usually negative for the reagent combinations recommended.

Any platelets or red cell debris included within the light scatter gate are usually negative for the reagents in the recommended panel. Debris, however, can produce increased nonspecific binding of the reagents, which can result in debris being counted with the positive lymphocyte subpopulations. This is especially true of aggregated platelets, which may nonspecifically bind antibodies and have light scattering properties similar to lymphocytes.

Correcting Lymphocyte Calculations

Corrections should be made to calculations of lymphocyte subsets, based on the purity of the light scatter gate.

1. Determine the percentage of each lymphocyte subset within the light scatter gate.
2. Determine the percentage of lymphocytes in the gate (purity) using CD45,CD14.
3. Calculate the corrected percentage of each lymphocyte subset in the gate by dividing the percentage of each lymphocyte subset (1) by the purity of the gate (2).

Quality control specifications for light scatter gate quality must be met for lymphocyte recovery and purity. With respect to lymphocyte recovery, a minimum of 90% of all lymphocytes should be included within the light scatter gate. It is important to realize that by setting smaller light scatter gates, different subsets of lymphocytes can be excluded from the analysis.

The purity of the lymphocyte gate (minimum 85%) is important in order to correct the subsequent data for nonlymphocytes included within the gate. With the reagent panel suggested for monitoring HIV disease, it is better to include more lymphocytes than to achieve maximum purity.

Absolute number calculations. A complete subset analysis must report the data both in terms of percentage positive lymphocyte subset, corrected for nonlymphocytes in the gate, and in absolute counts. The calculation of absolute counts, Equation IV.1, combines results from both the hematology and the flow cytometry laboratories.

The absolute number of positive lymphocytes per microliter is determined from hematology results by multiplying the automated lymphocyte differential percentage (% lymphocytes) by the white blood cell (WBC) count (cells/μL). This value is then multiplied by each of the corrected percentage positive lymphocyte subset values obtained from FCI results to obtain the absolute number of cells for each lymphocyte subset as follows:

Equation IV.1

$$\text{Absolute number of positive lymphocytes per } \mu\text{L} = \text{Corrected \% positive gated subset} \times \text{Number of lymphocytes per } \mu\text{L}$$

Example:

$$49\%\ \text{CD4+ T lymphocytes} \times 2{,}000/\mu\text{L} = 980\ \text{CD4+ T lymphocytes}/\mu\text{L}$$

Limitations of the system. The logical relationships for quality control may not always be met because of limitations of the specimen. For example, the lymphocyte concentration in the specimen may be too low. With lymphopenic specimens, it is necessary to count a larger number of total cells. If less than 2,500 lymphocytes are analyzed for each sample tube, there is increased variability in the determination because of the limit of the counting statistics.

A specimen may have a large number of dead cells or high levels of nonspecific staining. The logical relationships require complete separation of the major populations into discrete groups of cells. Therefore, changes in intensities of antigen staining or antigen density as well as an increased nonspecific staining will affect the discrimination of positive from negative cells.

Contamination of the light scatter gate by nonlymphocytes necessitates a significant correction when calculating the results. In order to make these calculations, it is essential to show that the light scatter pattern is identical for all samples within the specimen.

Alternate Calculation of Absolute Immunophenotyping Counts

Calculation of absolute CD4+ lymphocytes requires two separate leukocyte differentials, the first in setting a "lymphocyte" gate on the flow cytometer and the second in determining the percent lymphocytes by the hematology analyzer. This combination of two technologies to define lymphocytes introduces significant errors, especially in comparison of results from different sites. Occasionally, it is impossible to obtain an accurate differential count on the hematology analyzer.

An alternate method for calculating absolute counts requires FCI and the hematology WBC without a differential count. In this procedure, the CD3+,CD4+ cells (or other lymphoid subsets) are enumerated relative to the total leukocytes in the specimen and not as a percentage of lymphocytes alone. The denominator in the flow cytometric determination changes from positive cells per lymphocyte to positive cells per leukocyte. This eliminates the need to identify lymphocytes by light scattering and works well with the recommended HIV reagent panel. Other reagent combinations, which do not result in discrete positive populations, may not be used in this calculation without careful discrimination between specific and nonspecific staining.

In this procedure, the CD45, CD14 tube is used to identify the percent leukocytes in the specimen. Using the fluorescence histogram (as seen in Fig. 4.14), the debris is excluded and all CD45+ cells (leukocytes) are counted. No light scatter gate is used in this determination. Subsequent tubes are also analyzed without a light scatter gate. The discrete populations of lymphoid cells can be identified in most specimens, even with the inclusion of granulocytes. (Note: In some specimens the nonspecific staining from the granulocytes interferes with the accurate enumeration of the lymphocyte subset. A large light scatter gate including both the lymphocytes and monocytes can be used to exclude the granulocytes. Care should be taken to include all of the lymphoid subsets in the analysis. The number of lymphocyte positive cells should then be determined based on total cells counted, not on the gated events.)

The number of positive CD3+,CD4+ cells are divided by the number of total events analyzed. This number is then divided by the percent leukocytes in the specimen as determined in Tube #1 and multiplied by the WBC from the hematology analyzer to give absolute counts/μl by the following equations:

Equations IV.2

$$\frac{\text{Number CD3+/CD4+ cells}}{\text{Number total counts}} \div \frac{\text{Number leukocytes (CD45+)}}{\text{Total counts}} = \text{Lymphocyte subset/Leukocyte}$$

Then, using the WBC,

$$\frac{\text{Lymphocyte subset}}{\text{Leukocyte}} \times \text{WBC} = \text{Lymphocyte subset}/\mu\text{L}$$

For example, counting 10,000 total cells:

$$\frac{\text{1100 CD3+CD4+ cells}}{\text{10,000 Total counts}} \div \frac{\text{9000 CD45+ leukocytes}}{\text{10,000 Total counts}} \times \text{5000 Leukocytes}/\mu\text{L} = \text{611 CD3+,CD4+ cells}/\mu\text{L}$$

This approach may be used as an alternate calculation of absolute counts only if the laboratory has validated its accuracy in parallel with their usual calculations combining the hematology and flow cytometry differential.

CONCLUSIONS

Strict adherence to the CDC recommended immunophenotyping panel and the QA and QC parameters for which it was designed will allow more consistent CD4+ T-lymphocyte enumeration. These QC procedures should be used both to standardize CD4+ T-cell quantitation within every FCI laboratory and to ensure better agreement of quantitative immunophenotyping between laboratories.

FURTHER READINGS

Aledort LM, Hilgartner MW, Pike MC, Gjerset GF, Koerper MA, Lian EY, Lusher JM, Mosley JW. Variability in serial CD4 counts and relation to progression of HIV-I infection to AIDS in haemophilic patients. Transfusion Safety Study Group. *BMJ* 1992, 304:212–216.

Bofill M, Janossy G, Lee CA, MacDonald-Burns D, Phillips AN, Sabin C, Timms A, Johnson MA, Kernoff PB. Laboratory control values for CD4 and CD8 T lymphocytes. Implications for HIV-1 diagnosis. *Clin Exp Immunol* 1992, 88:243–252.

Burcham J, Marmor M, Dubin N, Tindall B, Cooper DA, Berry G, Penny R. CD4% is the best predictor of development of AIDS in a cohort of HIV-infected homosexual men. *AIDS* 1991, 5:365–372.

Centers for Disease Control. CDC 1993 revised classification system for HIV infection and expanded surveillance case definition for AIDS among adolescents and adults. *MMWR Morb Mortal Wkly Rep* 1992, 41(RR-17):1–19.

Centers for Disease Control. Guidelines for the performance of CD4+ T-cell determinations in persons with human immunodeficiency virus infections. *MMWR Morb Mortal Wkly Rep* 1992, 41(RR-8):1–17.

deShazo RD, Chadha N, Morgan JE, Shorty VJ, Rangan RS, Kalyanaraman VS, Hyslop N, Chapman DY, O'Dea S. Immunologic assessment of a cluster of asymptomatic HTLV-I infected individuals in New Orleans. *Am J Med* 1989, 86:65–70.

Fernandez-Druz E, Fernandez AM, Gutierrez C, Montes MG, de la Morena MT, Rodrkiguez-Villaneuva J, Longo N, Zabay JM. Progressive cellular immune impairment leading to development of AIDS: two-year prospective study of HIV infection in drug addicts. *Clin Exp Immunol* 1988, 72:190–195.

Giorgi JV. Characterization of T lymphocyte subset alterations by flow cytometry in HIV disease. *Ann N. Y. Acad Sci* 1993, 677:126–137.

Giorgi J, Hultin L. Lymphocyte subset alterations and immunophenotyping by flow cytometry in HIV disease. *Clinical Immunology Newsletter* 1990, 10:55–61.

Gorham ED, Garland FC, Mayers DL, Goforth RR, Brodine SK, Wiess PJ, McNally MS. CD4 lymphocyte counts within 24 months of human immunodeficiency virus seroconversion. Findings in the US Navy and Marine Corps. The Navy Retroviral Working Group. *Arch Intern Med* 1993, 153:869–876.

Haseltine WA. Molecular Biology of HIV-1. In Dalgleish AG, Weiss RA, eds. *AIDS and the New Viruses*. Academic Press: London, 1990.

Hoover DR, Graham NMH, Chen B, Taylor JMG, Phair J, Zhou SYJ, Munoz A. Effect of CD4+ cell count measurement variability of staging HIV-1 infection. *Journal of Acquired Immune Deficiency Syndromes* 1992, 5:794–802.

Hughes MD, Stein DS, Gundacker HM, Valentine FT, Phair JP, Volberding PA. Within-subject variation in CD4 lymphocyte count in asymptomatic human immunodeficiency virus infection: implications for patient monitoring. *J Infect Dis* 1994, 169:28–36.

Jackson A. Basic phenotyping of lymphocytes: selection and testing of reagents and interpretation of data. *Clinical Immunology Newsletter* 1990, 10:43–55.

Loken M, Brosnan J, Bach B, Ault K. Establishing optimal lymphocyte gates for immunophenotyping by flow cytometry. *Cytometry* 1990, 11:453–459.

Tindall B, Cooper DA, Donovan B, Penny R. Primary human immunodeficiency virus infection: clinical and serological aspects. In Sande MA, Volberding PA, eds. *The Medical Managment of AIDS*. WB Saunders, Philadelphia, 1988.

Yarchoan R, Venzon DJ, Pluda JM, Lietzau J, Wyvill KM, Tsiatis AA, Steinberg SM, Broder S. CD4 count and risk for death in patients infected with HIV receiving antiretroviral therapy. *Ann Intern Med* 1991, 115:184–189.

STUDY QUESTIONS

1. Describe in general terms how HIV infects cells and proliferates.
2. Draw a graph of the HIV clinical stages of infection (on the x-axis) versus relative levels of the following HIV indicators (on the y-axis): viral protein p24, IgG to HIV, and CD4+ T lymphocytes.
3. Name two of the types of human cells that are susceptible to attack by HIV.
4. Describe the factors used to classify HIV patients who are placed in clinical categories A, B, and C.
5. List the ranges in CD4+ T lymphocytes for laboratory categories 1, 2, and 3 in HIV-positive patients.
6. Draw a 3×3 grid showing how clinical categories (A, B, and C) and laboratory categories (1, 2, and 3) are used to stratify HIV-positive individuals and AIDS patients according to the CDC 1993 classification scheme. In each box indicate whether the individual is considered asymptomatic, symptomatic, or to be suffering from AIDS.

7. Give the CD designations and the reactivity of the two antibodies that are used to establish the lymphocyte light scatter gate.
8. Describe the difference in CD45 expression among lymphocytes, monocytes, and granulocytes.
9. Draw and label a plot of the major populations present in erythrocyte-lysed blood, with CD45-FITC (on the x-axis) and CD14-PE (on the y-axis).
10. Discuss the difference between lymphocyte purity and lymphocyte recovery in reference to the light scatter gate.
11. Name three subsets of lymphocytes that might be selectively eliminated from the data if the light scatter gate is set too small.
12. On a two-parameter fluorescence plot, draw a population displaying discrete antigen expression (on the x-axis) and a population with continuous antigen expression (on the y-axis).
13. Explain the significance of poor instrument sensitivity and improper spectral compensation on the results for a population that displays continuous antigen expression.
14. Explain how the results from multiple tubes containing CD3 can be used as a quality control check.
15. Explain how to use the double-negative cell population in each sample tube containing discrete populations as an internal negative control for nonspecific staining.
16. Name the nonlymphocyte population that could potentially interfere with obtaining the true CD4+ T-lymphocyte count.
17. Explain how to check for correct spectral compensation and nonspecific binding in the CD3,CD4 sample tube.
18. Explain how to assess within-specimen variation using CD3, and give the maximum recommended percentage difference between high and low values.
19. Give the lymphocyte summation equation and the acceptable range necessary for meeting this quality control check.
20. Compare the differences in relative percentages of CD4+ and CD8+ T lymphocytes from the time of HIV seroconversion, through the various clinical stages of HIV infection and AIDS.

V

Peripheral Blood Stem Cell Quantitation

INTRODUCTION

The strategy in high-dose chemotherapy as a treatment for cancer patients is to intensify the drug level to a concentration where all hematopoietic cells are effectively killed. For certain neoplastic cells this drug level will be sufficient to kill the tumor as well. Because the intensified chemotherapy dramatically reduces or eliminates all normal hematopoiesis, patients become neutropenic, thrombocytopenic, and anemic.

Without rescue with progenitor cells to replace the entire hematopoietic system, the patient would succumb to infectious diseases and hemorrhage. This period of aplasia must be minimized to reduce the acute life-threatening risks to the patient. The goal is to regenerate the blood cells rapidly once the tumor cells have been killed.

Transplantation by Peripheral Blood Stem Cells (PBSC)

Hematopoietic progenitor cells were originally obtained from bone marrow, hence the name bone marrow transplantation. Peripheral blood stem cell (PBSC) harvests, where the blood cell progenitors have been mobilized to migrate from bone marrow into blood, are rapidly replacing bone marrow as a source of progenitor cells in the treatment of cancer patients receiving intensified chemotherapy (Fig. 5.1).

Usually the stem cells come from the cancer patients themselves (*autologous transplants*), however the same procedure can be used

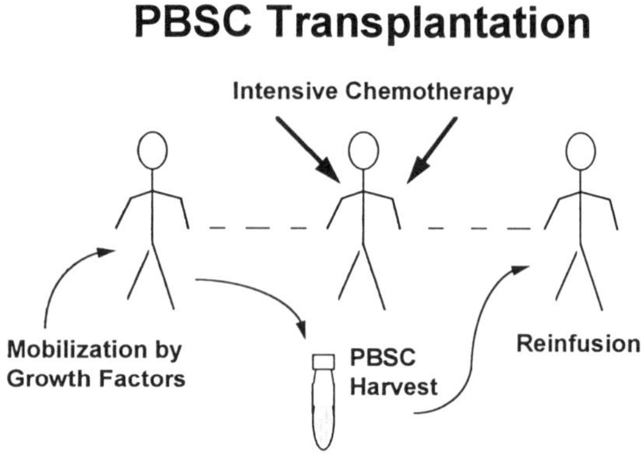

Fig. 5.1. Stem cell transplants permit high doses of chemotherapeutic drugs. The stem/progenitor cells are mobilized to migrate from bone marrow to blood by administration of hematopoietic growth factors. These stem cells are harvested and cryopreserved for later use. The patient is exposed to lethal doses of antitumor drugs, followed by the reinfusion of the harvested stem cells to repopulate all of the components of blood.

for related or unrelated donors (*allogeneic transplants*). This procedure is routinely performed for chronic and acute myelogenous leukemia, lymphoma, and disseminated breast cancer.

The definition of "time to successful engraftment" measures the time when the transplanted progenitor cells are proliferating sufficiently to obviate the need for granulocyte or platelet transfusions. Absolute neutrophil counts (ANC) must be greater than 500/μL and platelet counts greater than 20,000/μL for the graft to be considered successful. The average time to successful engraftment for a bone marrow transplant is between 14 and 21 days.

The major advantage of PBSC is more rapid engraftment as compared with an equivalent bone marrow transplant. Engraftment as early as day 10 is frequently achieved with PBSC. This is most likely a result of the higher yield of late progenitor cells available in the PBSC. This rapid engraftment means reduced infection rates for the patients and a decrease in hospital stays by 7–20 days.

For autologous transplants, it is believed that the blood has fewer contaminating tumor cells than bone marrow. The reinfusion of tumor cells back into the patient must be avoided, and stem cells collected from the blood would carry a lower risk for tumor recurrence.

Progenitor Cell Mobilization

The number of circulating lymphohematopoietic progenitor/stem cells is low in normal peripheral blood, but these cells can be increased in frequency by mobilization from the bone marrow into the blood by various growth factors. These growth factors can be increased endogenously as a result of cytotoxic drugs or exogenously through the administration of recombinant proteins such as granulocyte colony stimulating factor (G-CSF) or granulocyte-macrophage colony stimulating factor (GM-CSF). The maximum number of circulating progenitor/stem cells is reached 5–10 days after the induction procedure (with considerable patient-to-patient variation) so that timing of the leukapheresis becomes an important consideration.

Even with stimulation, progenitor/stem cells are found at extremely low frequencies within a heterogeneous group of other related cells. Rigorous quality control is essential during laboratory assessment of the progenitor/stem cell count to determine the frequency of these cells accurately. This number can be used to time the harvesting, to predict engraftment rate, and to identify high-risk patients.

Currently, timing of the apheresis is often done based on the white blood cell count (WBC) of the donor. When WBC is >10,000, apheresis is started. Few, if any, hospitals are using CD34 enumeration to define the starting point of the PBSC collection. The number of progenitor cells required for successful engraftment is quite consistent from laboratory to laboratory. The maximum rate of engraftment is found with 5×10^6 CD34+ cells/kg of recipient (Fig. 5.2). This translates to approximately 50×10^4 CFU-GM/kg. Transplanting more than this number does not speed the engraftment process. Patients with fewer than 5×10^6 CD34+ cells/kg are considered as high-risk patients.

Bioassays

The procedures used to quantify the progenitor/stem cells take two quite different forms: bioassays and immunofluorescence assays. The bioassay is based on the growth of hematopoietic colonies in tissue culture. The assays involve growing the cells for 1–2 weeks in a semisolid culture medium (e.g., low concentrations of methyl

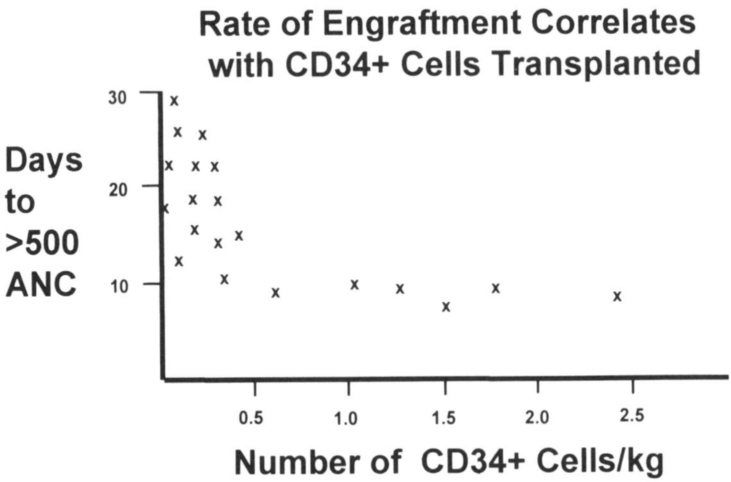

Fig. 5.2. Rate of engraftment increases with number of progenitor cells transplanted. The time to generate >500 neutrophils/μL following intensive chemotherapy drops to 10 days when 5×10^6 CD34+ cells/kg are reinfused (Bender, 1992). Scale on x-axis is $\times 10^7$.

cellulose) containing serum and/or albumin, other proteins, and added growth factors.

The CFU-GM assay predicts the time to recovery of neutrophils, which are the most important cells for the patient. Most laboratories use only the CFU-GM assay for quantifying progenitor cells since experience has shown that the other late progenitor cells are correlated to these numbers.

The routine use of this bioassay is limited by the time and cost required to perform the tests as well as by the expertise available. As with any bioassay, there is considerable variability in the procedures and the results. It is especially difficult to compare CFU-GM results between different laboratories. The recent availability of recombinant growth factors and commerical, lot-controlled culture media has reduced the potential variation between laboratories due to "technical" differences. This standardization reduces but does not eliminate the variability.

The total number of colonies counted is small, with a correspondingly large variation related to counting statistics. There can be significant differences between technicians in scoring the same plates. In addition, different bone marrow specimens may respond differently to the same mixture of growth factors used in the assay. This difference in "plating efficiency" may be related to prior therapy or biologic variability and is not yet understood.

The major limitation of the bioassay is the time required to obtain the results. Because of the culture time, these data cannot be used to make decisions as to when to begin apheresis, or when to stop because sufficient numbers of cells have been collected. The 1- to 2-week delay required for growing the cells means that the results can only be used in a correlative manner to identify high-risk patients who had limited numbers of progenitor cells collected.

Flow Cytometric Analysis

The identification of CD34+, which binds specifically to human lymphohematopoietic progenitor/stem cells, facilitates the enumeration of these immature cells. The CD34 antigen is a 115-kd glycoprotein expressed by approximately 1.5% of normal human bone marrow cells. All of the progenitor cells are included within this fraction, with the least mature of the progenitor cells expressing the greatest amount of the antigen. All lineages of blood cells have their origin in the CD34+ population, including lymphocytes, monocytes, neutrophils, eosinophils, basophils, erythrocytes, and megakaryocytes.

Several technical problems can introduce significant variability in the quantitation of the CD34+ cells in blood or bone marrow. This population is, at best, a very minor subpopulation of the cells being analyzed. During successful mobilization of the progenitor cells, the CD34+ population may reach 1% of the total nucleated cells. Since these cells are relatively infrequent, it is necessary to use as many different characteristics as possible to distinguish the true progenitor cells from the majority of other cells. The positive staining of CD34+ progenitor/stem cells must be distinguished from the nonspecific binding, autofluorescence, and nonviable cell fluorescence that may be in the specimen.

The combination of CD34 with other antibodies also helps to distinguish the specifically labeled progenitor cells from the majority of other cells. The intensity differences of CD45 expression have been used to highlight the CD34+ progenitor cells. Similarly, combinations of lineage-associated antigens have aided in distinguishing the more mature blood cells from the progenitor cells. Light-scattering characteristics also play an important role in focusing on the CD34+ cell population. The true progenitor/stem cells have low right-angle light scattering (a measure of internal cellular complexity) and low to intermediate amounts of forward light scattering (a measure of cell size).

The clinically important parameter from these results is the total number of CD34+ cells available to transplant back to the patient. This total is obtained from Equation V.1:

Equation V.1

$$\frac{\text{CD34+ cells}}{\text{Total nucleated cells}} \times \text{WBC} \times \text{volume} = \text{Total CD34+ cells}$$

The first term is obtained from flow cytometric results, which express the CD34+ cells as a percentage of the total nucleated cells. The WBC is determined by standard cell counters and usually is performed at the collection site. The volume of the apheresis product is also determined following collection. Therefore, the CD34 quantitation is similar to CD4 determination in that the clinical result uses data obtained from different instruments, often at different sites. This emphasizes the importance of proper quality control being performed at every step of the procedure.

IMMUNOPHENOTYPING PANEL

The accurate detection of the infrequent number of CD34+ cells in blood (or bone marrow) requires careful discrimination of the specifically labeled cells from other cells that may improperly be counted as positive. The inclusion of only CD34+ cells will affect the numerator of Equation V.1.

The specifically labeled progenitor cells have lower autofluorescence than do monocytes or neutrophils. Cells of different lineages have differing propensities to bind antibodies nonspecifically. Dead cells also nonspecifically bind antibodies, thereby providing further complications in accurately enumerating the infrequent population.

The selection of the antibody and chromophore labeling can affect the resolution of the progenitor cells from the background of normal cells. At least three different epitopes have been described on the CD34 antigen. To date, the antibodies identifying the chymopapain-resistant epitope provide the best resolution when directly conjugated with the brightest chromophore, phycoerythrin. Some direct conjugates of antibodies to CD34 with FITC result in populations that are continuous with the unstained cells, and reproducible setting of the cursor between positive and negative is then difficult. Many antibodies that bind to the chymopapain-sensitive epitope of CD34 antigen cannot be directly conjugated and must be used indirectly with a second antibody to identify the cells binding the CD34. An F(ab')2 fragment for the secondary antibody is required if antisera are used for the second step since this preparation reduces nonspecific binding.

The careful titration of the negative control is also crucial since this defines the position of the boundary between negative and positive cells. Small positional changes in this setting can dramatically change the number of CD34 cells identified as positive. This titration must not be performed on the total cell population since monocytes and neutrophils have higher autofluorescence than do progenitor cells. This distinction is important for proper establishment of the boundary between positive and negative cells.

The panel used to enumerate the CD34+ cells, therefore, must take all of these factors into account:

1. Detection of viable cells only
2. Exclusion of committed myeloid cells
3. Maximization of the CD34 signal
4. Titration of negative control

All of these important features can be included in two tubes using three-color analysis in a recommended panel for CD34 enumeration:

TABLE V.1. Reagent Panel

Tube number	Green	Orange	Red
1	CD14	IgG1-PE	7-AAD
2	CD14	CD34-PE	7-AAD

The accurate enumeration of the CD34+ cells requires using all five parameters on the flow cytometer to exclude all nonprogenitor staining by the monoclonal antibody.

Discrimination of Viable Cells Using 7-AAD

The nucleic acid-binding dye, 7-amino-actinomycin D (7-AAD), is similar to propidium iodide in its ability to stain dead cells, except the emission occurs in the far red rather than in the orange part of the visible spectrum. This molecule is large, with two amino acid side chains that inhibit it from crossing intact plasma membranes. Therefore, 7-AAD is excluded from viable cells but freely passes into dead cells, which have their membranes disrupted.

Upon binding of the 7-AAD to DNA the fluorescence quantum yield is greatly increased as compared with the molecule in solution. The fluorescence coming from nonviable cells is easily distinguished from that from the unstained, live cells. This fluorescence is detected in the red channel of the flow cytometer, with no emission detected in either the green or orange channels. This means that two-color immunofluorescence can be obtained while performing live-dead discrimination.

The first step in CD34 analysis is the identification of the live cells (Fig. 5.3). All subsequent immunofluorescence analyses are performed on only these cells.

The importance of this step is illustrated in Figure 5.3 where the CD34 staining of the viable cells overlaps with the nonspecific staining of the dead cells.

The dead cells are not completely ignored, since they must be counted to be included into the determination of total cell number (Eq. V.1). All CD34 enumerations must be based on total nucleated cells, including both live and dead cells. Most cell counters used to determine WBC cannot distinguish between live and dead cells since they usually count nuclei, not cells. Therefore, the proportion of CD34+ cells must be calculated with a denominator of the total nucleated cells (both live and dead).

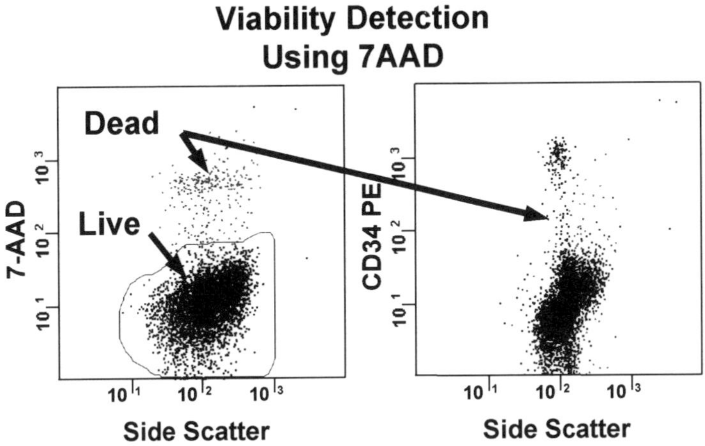

Fig. 5.3. 7-AAD to exclude dead cells. Dead cells from a PBSC harvest stain with 7-amino-actinomycin D. These dead cells can be gated out of the subsequent immunofluorescence analysis.

Excluding Committed Myeloid Cells

The appearance and disappearance of cell surface antigens and changes in light scattering characteristics are highly regulated events in hematopoietic development. These changes in cellular characteristics have been used to mark the progression of myeloid cells from progenitor cells to the mature, functional cells found in blood. These cellular changes can be used to focus on the progenitor cells while discriminating against cells that will confound the analysis.

The late myeloid cells can be eliminated by a combination of cell surface antigen detection and right angle light scatter. CD14 is an antigen that is expressed during monocyte development and reaches high levels in the mature monocyte. This antigen appears after CD34 is lost from the myeloid cells. Thus, during normal development, no cells coexpress CD34 and CD14. This antigen can then be used to eliminate cells committed to the monocyte lineage from the CD34 analysis.

Similarly, neutrophils and eosinophils dramatically increase their right angle light scatter signal early in their maturation as a result of the formation of primary and secondary granules specific to these cells. True progenitor cells do not have these granules, and as a result, they have low right-angle light scatter. This means that light scatter can be used as an independent means to gate out cells committed to the neutrophil and eosinophil lineages in a similar manner to gating on the expression of a specific antigen.

The discrimination against the mature myeloid cells is illustrated in Figure 5.4. The CD14-FITC is correlated with right angle light scatter. The maturing monocytes and neutrophils can be easily distinguished from the other cells based on these two parameters. The cells in the lower left hand corner include lymphocytes, nucleated erythrocytes, and progenitor cells. The autofluorescence and nonspecific binding of these cells is significantly less than that for the monocytes and neutrophils. Further analysis is performed on the cell population excluding myeloid cells.

Isotype Control

The accurate quantitation of CD34 requires the proper establishment of the boundary between negative and positive cells based

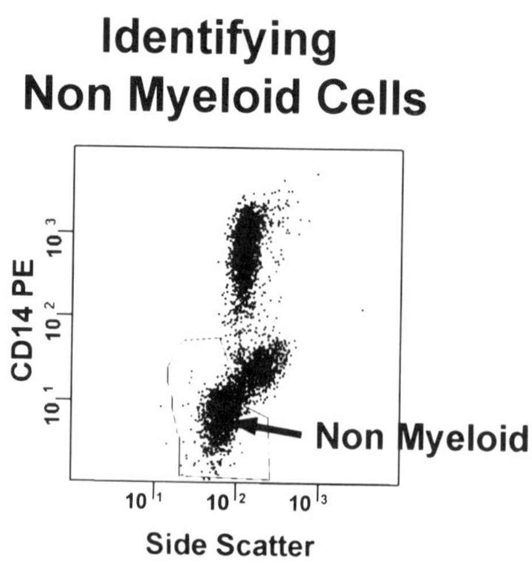

Fig. 5.4. CD14 and right angle light scatter of progenitor cells. Progenitor cells do not express CD14 and have low right-angle light scatter. The nonmyeloid cells contain the CD34+ population.

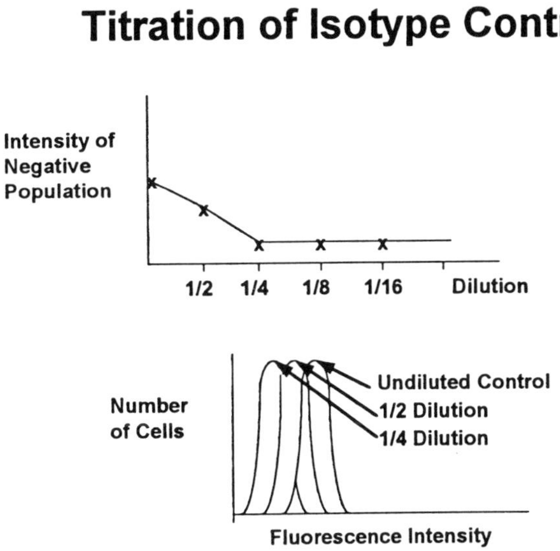

Fig. 5.5. Titration of isotype controls for CD34 analysis. Isotype controls must be titrated to discriminate CD34+ cells from CD34− cells. The viable, nonmyeloid cells (see Fig. 5.3 and 5.4) are treated with different dilutions of the proper isotype control. The correct dilution of the isotype control exhibits identical fluorescence to unstained cells gated in the same manner.

on a reference population. The proper reference cell population is viable (7-AAD−), CD14− and has low right-angle light scattering. This population should be exposed to a monoclonal antibody isotype control that does not bind to the cells at reagent concentrations comparable to those used for CD34 enumeration.

The isotype-matched control must be titered in a manner similar to that for antibodies with specific reactivities in order to define the appropriate concentration for use. In this case the negative cell population must be matched to the unstained, autofluorescence control (Fig. 5.5).

A plot of the fluorescence intensity of the negative cells versus dilution will show the proper concentration for use in the assay. All of the cells in the isotype control must be included in the unreactive population with the same fluorescence distribution as unstained cells.

CD34 Discrimination

Once the gates have been set to exclude cells that may confound the analysis (7-AAD, CD14, and right angle light scatter), the CD34 staining can be determined. The population of CD34+ cells identified by the phycoerythrin-coupled antibody is clearly resolved from the background. Frequently, the positive cells comprise two populations, bright CD34+ cells and dim CD34+ cells (Fig. 5.6).

The inclusion of the dim population in the CD34+ enumeration has not been rigorously investigated. These dim CD34+ cells are not separated from the background using antibodies recognizing the chymopapain-sensitive epitope. Therefore, in studies using these antibodies, this population is not included. Likewise, improper titration of the isotype control can raise the boundary between negative and positive cells to a level where these dim CD34+ cells are not included in the count.

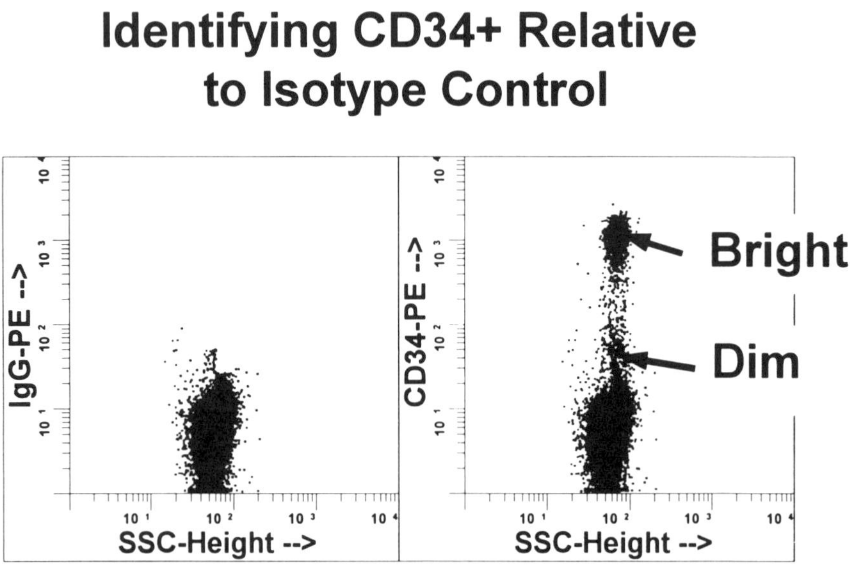

Fig. 5.6. PBSC stained with titered IgG-PE or with CD34-PE. The cell population must be gated as described in Figures 5.3 and 5.4. Intensity of CD34 expression often identifies both bright and dim populations.

Sorting studies have shown that CFU-GM colonies can be grown from the dim cells; thus they should be included in the quantitation of progenitors. The dim CD34 population can be 50% or more of the entire CD34+ cells. Inclusion or exclusion of this population can significantly change the total number of CD34+ cells quantified.

With careful attention to quality control procedures, highly reproducible numbers can be obtained for infrequent populations such as CD34+. This, however, requires routine boundary setting and careful definition of the negative control.

Number of Cells to Analyze

The CD34 population in a PBSC harvest is often under 1% of the total nucleated cells. In order to achieve adequate precision of the data, increased numbers of cells must be analyzed when compared with lymphocyte subpopulation analyses. This total number of cells processed is often a balance between the proportion of positive cells in a specimen and the practical time and computer storage space required to analyze large numbers of cells. Assuming random counting errors and Poisson statistics, the standard deviation of a determination is approximately the square root of the number of events counted. For 100 positive cells, the standard deviation is approximately ± 10. If higher precision is required for the determination, more events must be counted.

Flow cytometric analyses have routinely been performed on 10,000 cells. If the CD34 proportion is 1%, a total of 100 CD34+ cells will be identified. Since PBSC harvests often have fewer CD34+ cells, more cells must be analyzed to attain the same counting precision (Table V.2). Analysis of 50,000 cells will count 100 positive events when the CD34 proportion is 0.2%.

TABLE V.2. Total Cells Counted at Different CD34 Proportions

	Cells analyzed		
% CD34+	10,000	30,000	50,000
1.0	100	300	500
0.5	50	150	250
0.2	20	60	100

For the enumeration of CD34+ cells in peripheral blood, the expected range is 0.1–5.0%. With these proportions in mind, the numbers of cells routinely analyzed by flow cytometry should be between 30,000 and 50,000 events. Longer analyses may give slightly improved precision; however, the time required to collect the data and to perform reanalysis is not routinely justified.

If there is a need to quantify the CD34+ cells at proportions of 0.1% and below, it is recommended that 100,000 or more cells be analyzed.

Calculating Total CD34+ Cells

As in the analysis of CD4+ lymphocytes, accurate enumeration of CD34+ cells requires including only progenitors in the determination (purity) and including all of the precursor cells in the sample (recovery). The exclusion of dead and lineage-committed cells is required to accurately determine the numerator of Equation V.1.

The number of CD34+ cells must be expressed per nucleated cell in the specimen. Hematology instruments used to determine the WBC of the apheresis product count all nuclei, including dead cells and nucleated erythroid cells. A corresponding cell analysis by flow cytometry can be made by also including these cells in the determination. All nucleated cells include all 7-AAD+ cells, all CD14+ cells, and all cells with high right-angle light scatter. The number of CD34+ cells per WBC can then be determined by the following Equation V.2:

Equation V.2

$$\frac{(\text{Number of CD34+ cells}) - (\text{number of positive cells in isotype control})}{(\text{Number of nucleated cells in CD34 sample})} = \text{CD34+ cells/WBC}$$

The easiest procedure to determine CD34+ cells per nucleated cell is to collect the 30,000–50,000 cells in list mode. The *number* of cells in each group rather than the *percentage* should be determined, since different gates are used to identify the CD34+ cells and the total nucleated cells.

The number of nucleated cells (excluding only platelets and mature red blood cells) in this data set should be calculated by reanalyzing the list mode data. This includes 7-AAD+ and 7-AAD− cells, CD14+ cells, and high and low right-angle light scatter cells; the number of nucleated cells should approximate the total number of events in the list. This number becomes the denominator in Equation V.2.

A second analysis of the same list mode data, gating out the nonviable cells, the CD14+ cells and the cells with high right-angle light scatter, is used to determine the number of CD34+ cells. Usually this number will be between 100 and 1000, depending on the frequency of the CD34+ cells. This number should be entered into Equation V.2.

The isotype control should then be analyzed with the exact gates used for the CD34+ determination, to correct for the number of nonspecific stained cells expected in the CD34 enumeration. This isotype control sample must have the same total number of cells in the list mode as are present in the CD34 list.

The number of CD34+ cells, corrected for the expected number of nonspecific stained cells (from the isotype control), divided by the total nucleated cells in the list, provides the CD34+ cells per nucleated cell. If the isotype control has not been properly titered, it is possible to obtain fewer cells in the CD34+ fraction than are identified by the negative, isotype control.

The total number of CD34+ cells in each apheresis product is determined from Equation V.3 by multiplying the CD34+ cells per nucleated cell times the WBC and volume of the cell harvest:

Equation V.3

$$\frac{\text{CD34+ cells}}{\text{Nucleated cell}} \times \text{WBC} \times \text{volume} = \text{Total CD34+ cells}$$

This number, divided by the weight of the recipient in kilograms, has been used to identify those patients who will engraft rapidly. The patients receiving $>5 \times 10^6$ CD34+ cells/kg will usually engraft with >500 neutrophils by day 10–14. Patients with less than this number often continue apheresis for more than one cycle to increase the number of progenitors. Patients with good response to growth factors can achieve this level in a single apheresis session.

CONCLUSIONS

The ability to intensify chemotherapy has significantly increased survival rates for patients with disseminated cancer. Levels of cytotoxic drugs that will kill the tumor can now be attained, and the patient can be rescued by progenitor cell transplant. The frequency of bone marrow transplantation procedures has remained constant over the last few years while an exponential increase in PBSC has been reported. Techniques have been improved to provide such treatment on an outpatient basis so that hospitalization occurs only if complications are evident.

The identification of high-risk patients, enhanced mobilization of progenitor cells resulting in a single apheresis harvest, detection of residual tumor cells in the blood, and hematopoietic growth factor administration following transplant will all facilitate the utilization of this procedure for a larger group of patients. Accurate enumeration of the progenitor cells is key to the success of this treatment modality.

FURTHER READINGS

Andrews RG, Singer JW, Bernsteir ID. Precursors of colony-forming cells in humans can be distinguished from colony-forming cells by expression of CD33 and CD34 antigens and light scattering properties. *J Exp Med* 1989, 169:1721.

Bender JG, To LB, Williams S, Schwartzberg LS. Defining a therapeutic dose of peripheral blood stem cells. *Journal of Hematotherapy* 1992, 1:329.

Bender JG, Van Eps D. Flow cytometric analysis of human bone marrow. In Gross SR, Gee AP, Worthington-White D, eds. *Bone Marrow Purging and Processing*. Alan R. Liss, New York, 1990, pp 137–154.

Bregni M, Siena S, Magni M, et al. Circulating hematopoietic progenitors mobilized by cancer chemotherapy and by rhGM-CSF in the treatment of high-grade non-Hodgkin's lymphoma. *Leukemia* 1991, 5(suppl 1):123.

Civin CI, Loken MR. Cell surface antigens on human marrow cells: dissection of hematopoietic development using monoclonal antibodies and multiparameter flow cytometry. *Int J Cell Cloning* 1987, 5:1.

Civin CI, Strauss LC, Fackler MJ, Trischmann TM, Wiley JM, Loken MR. Positive stem cell selection: basic science. Leukemia. Gross SR, Gee AP, Worthington-White D, eds. In *Bone Marrow Purging and Processing*. Alan R. Liss, New York, 1990, pp 387–402.

Civin CI, Strauss LC, Brovall C, et al. Antigenic analysis of hematopoiesis, III. A hematopoietic cell surface antigen defined by a monoclonal antibody raised against KG-1a cells. *J Immunol* 1984, 133:157.

Loken MR, Civin CI, Shah VO, et al. Characterization of erythroid, lymphoid, and monomyeloid lineages in normal human bone marrow. Laerum OD, Bjerknes R, eds. In *Flow Cytometry in Hematology*. Acadmic Press, London, 1992, pp 31–42.

Rabinovitch PS, Torres RM, Engel D. Simultaneous cell cycle analysis and two-color surface immunofluorescence using 7-amino-actinomycin D and single laser excitation: applications to study of cell activation and the cell cycle of murine Ly-1 B cells. *J Immunol* 1986, 136:2769–2775.

Siena S, Bregni M, Brando B, et al. Flow cytometry for clinical estimation of circulating hematopoietic progenitors for autologous transplanation in cancer patients. *Blood* 1991, 77:400.

Siena S. Bregni M, Brando B, et al. Flow cytometry for clinical estimation of circulating hematopoietic progenitors for autologous transplanation: comparative analysis of different CD34 monoclonal antibodies. *Haematologica* 1991, 76:330.

Socinski MA, Elias A, Schnipper L, et al. Granulocyte-macrophage colony stimulating factor expands the circulating hematopoietic progenitor cell compartment in man. *Lancet* 1988, 1:1194.

Terstappen LWMM, Huang S, Safford M, Lansdorp PM, Loken MR. Sequential generations of hematopoietic colonies derived from single nonlineage-committed CD34+CD38− progenitor cells. *Blood* 1991, 77:1218–1227.

Terstappen LWMM, Loken MR. Multi-dimensional flow cytometric characterization of myeloid maturation in normal bone marrow and acute myeloid leukemia. In Burger G, Oberholzer M, Vooijs GP, eds. *Advances in Analytical Cellular Pathology*. Excerpta Medica, Amsterdam, 1990, pp 209–210.

Trischmann TM, Schepers KG, Civin CI. Measurement of CD34+ cells in bone marrow by flow cytometry. *Journal of Hematotherapy* 1993, 2:305–313.

STUDY QUESTIONS

1. What are progenitor cells and how are they identified?
2. Explain why rescue with progenitor cells is an important part of high-dose chemotherapy protocols.
3. Differentiate between *autologous* transplantation and *allogeneic* transplantation.
4. List at least three diseases for which PBSC transplantation has become a routine procedure.

5. What characterizes a successful progenitor cell transplant?
6. What are the advantages of PBSC transplantation compared to bone marrow transplantation?
7. Describe how progenitor cell mobilization is accomplished.
8. How many progenitor cells are needed for a successful engraftment?
9. What are the disadvantages of the CFU-GM bioassay?
10. Why is CD34+ cell quantitation considered a rare event analysis?
11. Describe the light-scattering and CD45-staining characteristics of CD34+ progenitor cells.
12. Explain why all CD34 monoclonal antibodies may not have the same specificity.
13. Why must the negative control be titered along with the CD34 monoclonal antibody for this test?
14. Name at least three requirements for the CD34+ flow cytometric immunophenotyping panel.
15. Explain the five parameters of data collected in the progenitor cell immunophenotyping panel.
16. Why is 7-AAD an important reagent in the CD34 immunophenotyping panel?
17. How are mature monocytes, neutrophils, and eosinophils excluded from this analysis?
18. Describe how to set the boundary between negative and positive cells in the progenitor cell immunophenotyping panel.
19. Explain why the number of peripheral blood cells analyzed for CD34 by FCI may be different from patient to patient.
20. Why are the absolute count calculations for CD34+ cells corrected for the isotype control?

VI

Variability in Immunophenotyping

INTRODUCTION

How accurate are the immunophenotyping laboratory measurements of lymphocyte subsets? There is no standard for measuring the accuracy of cell quantitation, although interlaboratory surveys can provide comparative information about CD4+ cell measurement. Unfortunately, the agreement between laboratories has been disappointing, according to the College of American Pathologists (CAP) and the AIDS Clinical Trial Group (ACTG), prompting the recent National Committee for Clinical Laboratory Standards (NCCLS) and CDC guidelines for the performance of flow cytometric immunophenotyping. No surveys or guidelines are available for quantitation of CD34 cells.

There are several obvious reasons for the interlaboratory variability, including the lack of standard materials and training programs, the high direct cost of testing, and the unavailability of true proficiency testing. The lack of compliance by clinical laboratories is often based solely on cost, since the additional measurements recommended for quality control purposes can significantly increase the cost of testing, a cost not always fully understood by practicing physicians or reimbursement agencies.

In 1993, the Revised CDC Guidelines for Human Immunodeficiency Virus Infection and Disease Surveillance in Adults and Adolescents combined clinical parameters with the laboratory quantitation of CD4+ T lymphocytes to give a new definition of AIDS. By this redefinition, the flow cytometric laboratory measurement of CD4+ T lymphocytes in peripheral blood became the critical assessment for the management of individuals who are infected with HIV. Similarly, the total number of progenitor cells for transplant patients is a critical determination that may affect subsequent care. Variability in these measurements now becomes a central issue for patient management.

What are the inherent problems in lymphocyte subset quantitation? As in many clinical tests, problems may arise in both analytic and biologic areas. While the following discussion may be more relevant to the quantitation of CD4+ T lymphocytes, issues of analytic and biologic variability affect peripheral blood stem cell quantitation as well.

ANALYTIC VARIABILITY

Nearly every stage and every variable, from drawing the patient's blood to the reporting of results, can influence CD4 or CD34 cell quantitation. Type of anticoagulant, storage temperature, and time before processing can all affect cell viability and thus influence cell recovery. Since CD4+ T cells are measured as a percentage of lymphocytes, the loss of any lymphocytes can influence the CD4 result.

The absolute CD4+ T-lymphocyte count is a product of the percent of CD3+CD4+ lymphocytes measured by flow cytometry, and the WBC and percent lymphocytes from hematology analysis. Likewise, both methodologies are used for CD34 quantitation. The combination of two methodologies increases the variability of CD4 and CD34 results.

Sample processing is extremely important for cell recovery. Current procedures recommend whole blood staining, with subsequent lysis of erythrocytes. However, variability can occur when erythrocytes are not completely lysed. Laboratories still using the ficoll/hypaque density gradient separation to isolate peripheral blood mononuclear cells are unable to control for cell loss during this process. Protocols that require isolated cells to be cryopreserved with DMSO (dimethyl sulfoxide) for later analysis cannot control for cell loss during either the freezing or thawing stages.

Although sample processing is thought by Gelman and coworkers to represent the majority of analytic variability in FCI, interlaboratory studies show that differences in monoclonal specificity, variables in instrumentation, and, more recently, changes in gating procedures can significantly affect CD4 measurements. Gating can be automatic or manual, based on light scatter parameters or on fluorescence, or both.

Today's measurement of CD4+ T lymphocytes uses at least two-colors for immunophenotyping and requires that these cells be defined as CD3+CD4+. In the past, CD4+ cells were not necessarily characterized as T cells (CD3) or even lymphocytes! In a similar manner, the discrimination between specific and nonspecific staining for CD34 dramatically changes the results.

The greatest source of variability in calculating absolute CD4 counts is in the hematology measurement of WBC and lymphocyte percentages. Proficiency testing indicates that leukocyte counts have a CV of 4.6%, and percentage lymphocytes a CV of 16.9%. Some laboratories recommend the reporting of CD4 results as percentages, in order to minimize such outside variables.

A recent report by Hughes and coworkers characterized the "within subject" variation in CD4+ lymphocyte counts of 1020 untreated patients with asymptomatic HIV infection to be 25%. Specimens taken 8 weeks apart were compared for CD4+ cell count (Fig. 6.1).

Although the authors concluded that both analytic and biologic variability contributed to these CD4 fluctuations, they described the coefficients of variation of the WBC, the lymphocyte percent, and the CD4 percent to be 15%, 21%, and 18%, respectively. One must therefore recognize that the hematology results alone could account for the majority of patient variability in this cohort.

In spite of the inherent analytic variability of flow cytometry, recent trends show that interlaboratory variability is improving. CDC guidelines for CD4 immunophenotyping have already been revised, in response to clinical laboratory practice, and national training programs have been created to facilitate standardization of procedures. However, even if analytic variability is minimal, biologic variability must always be considered in the interpretation of immunophenotyping values.

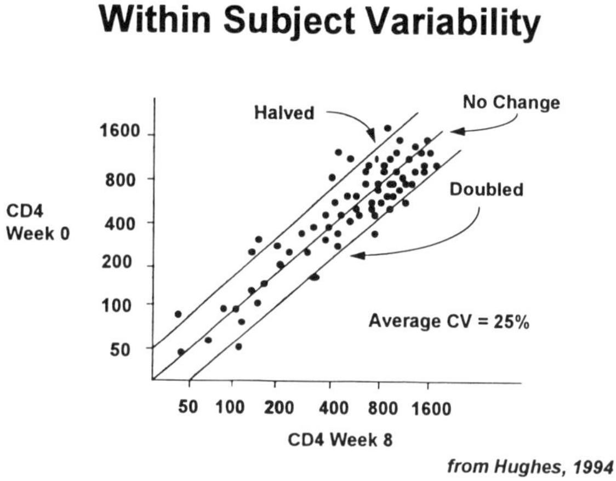

Fig. 6.1. Analytic variability of CD4. Specimens taken 8 weeks apart from a large cohort of asymptomatic HIV-seropositive individuals showed a mean CV of 25% (results expressed as cells/μL). Although biologic variability could contribute to these fluctuations, the authors describe sources of analytic variation that could account for the observed variability.

EVOLUTION OF IMMUNOPHENOTYPING METHODOLOGY

One of the greatest obstacles to consensus in measuring CD4+ T lymphocytes has been the rapid evolution of measurement techniques over the past decade. Early studies in CD4 cell measurement were performed microscopically. Accuracy of absolute counts assessed by microscopy may be challenged due to the low percentage of circulating cells that are actually counted on a slide.

When flow cytometric immunophenotyping (FCI) became available, precision improved as thousands of cells could be evaluated within minutes. The original techniques relied on isolating mononuclear cells by a density gradient technique, ficoll-hypaque, and purifying T cells by their binding to sheep red blood cells. The red cells were lysed and lymphocytes were further identified by specific monoclonal antibody binding followed by a secondary fluorescent-tagged antiserum. These isolation techniques could not evaluate cell recovery and therefore, in the strictest sense, should not be used to judge absolute CD4+ T-cell numbers.

Currently, CDC guidelines recommend that CD4+ T-lymphocyte immunophenotyping be performed by dual-color FCI using whole blood methodology. This technique has the advantage of more reproducible cell recovery and of improved precision in identifying CD4+ T lymphocytes for quantitation. Clearly, it is impossible to directly compare data from older microscopic studies with FCI studies. It is also inappropriate to compare indirect immunophenotyping methods with two-color direct FCI. However, in an effort to provide as much information as possible, whether quantitative or qualitative, regarding factors or circumstances that may influence CD4 measurements, the following discussion is presented.

These older data must be accepted with the caveat that cell quantitation on ficoll-separated peripheral blood cells may not be accurate, but may provide information regarding relative proportions of cells. More importantly, the identity of the cell population being measured must be considered, especially in light of today's definition of CD4+ T lymphocytes as CD3+CD4+ cells. Unless otherwise noted, the FCI analysis described in the following reports is a single-parameter evaluation of CD4, not conforming to today's definition. The general concepts can also be applied to CD34 enumeration as well.

SPECIMEN QUALITY

The quality of a clinical laboratory specimen dictates the quality of the analytic result. Inappropriate or abused specimens may increase variability in FCI by creating inappropriate light scatter patterns through cell clumping, cell death, and debris.

Anticoagulants

In order to provide a cell suspension for FCI, clinical specimens must be anticoagulated; however, not all anticoagulants are equally appropriate for FCI. For CD4+ T-lymphocyte immunophenotyping in peripheral blood, one has several choices: ethylenediaminetetraacetic acid (EDTA) may be used if the specimen will be processed in less than 30 hours after draw. If the delay is more than 30 hours, however, sodium heparin or acid citrate dextrose (ACD) are the anticoagulants of choice because they preserve cell morphology more effectively. For peripheral blood stem cell (PBSC) quantitation, specimens must be drawn into ACD. Bone marrow specimens are generally heparinized.

All anticoagulants require complete mixing with the blood specimen to prevent clotting. Specimens with clots, even small ones, can produce variable results and can also block the flow cell.

Time

As soon as blood leaves the body, cells start to die. Blood specimens should be transported to the laboratory as soon as possible after collection. Although guidelines recommend that EDTA blood be analyzed for FCI prior to 30 hours, many laboratories receive blood within 6 hours of draw and others do not accept EDTA blood after 24 hours. The least variable results come from fresh specimens, wherein light scatter patterns are most easily discriminated.

Temperature

Specimen quality is also directly related to its storage and transport temperature. Optimal temperature for peripheral blood specimens is room temperature, between 18 and 22°C. Specimens should never be heated or chilled and, of course, never frozen. Transportation of specimens to the lab should be as direct as possible, with safety and temperature controls. Tubes of blood should be packed in a plastic, leak-proof container, like a zip-lock bag, and placed inside a rigid container for transport. A second container, usually insulated against extreme temperature and packed with absorbent material, holds the transport container.

SAMPLE PROCESSING

On arrival at the lab, all specimens must be checked for acceptability. It is critical that the date and time of draw be included on the tube or on accompanying documents so that the laboratory can determine if the specimen meets the criteria for immunophenotyping. In addition, each tube should be monitored for temperature, assessing whether the specimen is hot or cold to the touch. A visual inspection for hemolysis and for cell clotting should also be made and documented.

Those specimens that satisfy the laboratory's QA standards on receipt are processed for immunophenotyping. During processing, variability can be introduced by imprecise pipeting devices, cell loss during centrifugation or cell isolation on density gradients, and effects of the erythrocyte lysis method.

ANALYSIS PROCEDURES

Variability can also be introduced into immunophenotyping during analysis on the flow cytometer. For example, gating procedures must be standardized and controlled in order to achieve reproducible results. The CDC recommendations for quantitative CD4+ T lymphocytes include requirements for both lymphocyte recovery and purity of the analysis gate. Recovery should be at least 90% and purity should be at least 85%.

When determining progenitor cell frequencies, it is necessary to follow the identical procedures for gating out dead cells and committed myeloid cells for each specimen. The frequency of CD34+ cells is usually under 1%, so small populations of cells that nonspecifically bind the antibody can cause wide fluctuations in the number of CD34+ cells identified. The inclusion of a 1% nonspecific stained population with the progenitor cells can cause a 100% error in total CD34+ cells counted. Careful attention to the proper detection of the CD34+ cells reduces the variability to a minimum.

Measuring Analytic Variability

As a component of its quality assurance program, each laboratory should characterize its own analytic performance and establish baseline operating standards. For example, the laboratory should have data on two types of reproducibility: (1) within-sample reproducibility as a measure of instrumentation and (2) between-sample reproducibility as a measure of sample preparation methodology.

Instrument precision and reproducibility. Replicate measurements of a single sample on the same flow cytometer provide a measure of that instrument's precision and reproducibility (Fig. 6.2).

The variation observed in 10 replicate determinations of the same sample is unaffected by sample preparation. The greatest source of variation in this test is the instrument itself. Ideally, the standard

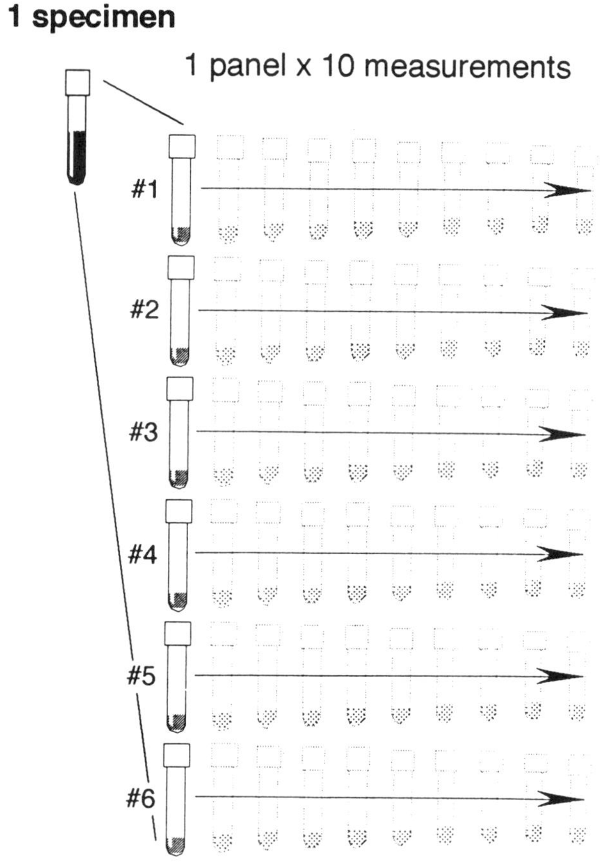

Fig. 6.2. Measuring instrument reproducibility. Samples set up from a single blood specimen should be run on the flow cytometer 10 times. The results should be compared and the standard deviation calculated. This value estimates the precision of the instrument itself.

deviations obtained in this test should be less than 1% and should correspond to the maximum precision obtainable with the instrument.

Procedural precision and reproducibility. Measurements of replicate panels prepared from the same specimen give a realistic indication of the reliability or precision of determinations (Fig. 6.3). Since sample processing includes many potential sources of variability, the procedural reproducibility will not be less than the instrument

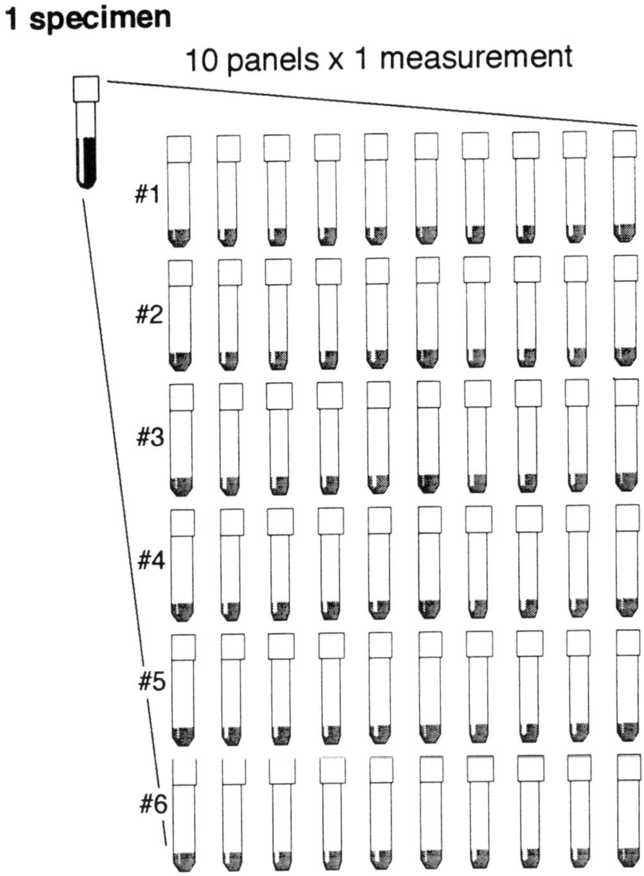

Fig. 6.3. Measuring procedural reproducibility. Set up 10 replicate panels from the same blood specimen and run each on the flow cytometer. Calculate the mean, standard deviation, and confidence interval for each analyte.

reproducibility. Sample preparation now adds a second measure of variability to instrument variation.

Data may be charted for visual review (Fig. 6.4) or analyzed in a spreadsheet format. For each parameter measured, the associated standard deviation can be used to compute a confidence interval for a single measurement. For example, if the calculated standard deviation (SD) for percent CD4 is 0.8, a 95% confidence interval for a single determination would be the measured value ±2 SD—in

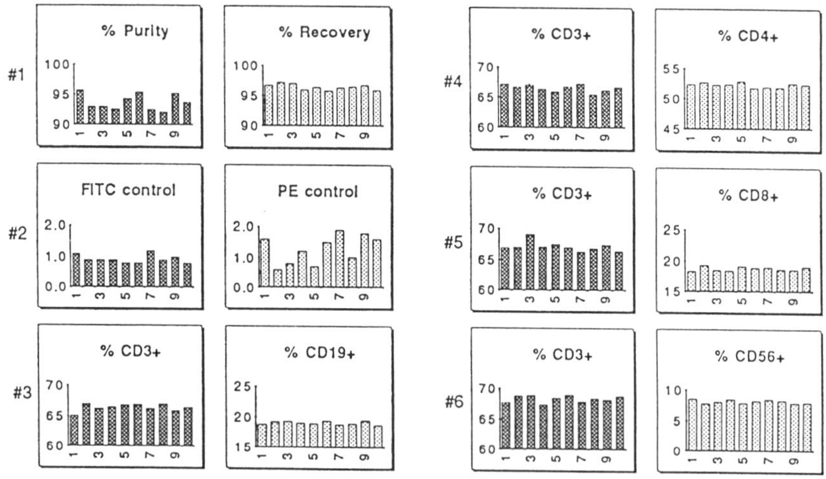

Fig. 6.4. Charting reproducibility data by analyte. For each component of an immunophenotyping panel, calculate the 95% confidence interval. This calculation should be made for instrument variability and also for procedural variability.

this case, ±1.6%. In other words, there is only a 5% chance (1 in 20) that the patient's actual percent CD4 lies outside this 3.2% range. This assessment can be used to set the limit of biologic significance that can be inferred from the data.

Once a laboratory has characterized its variability, this information can be used in several ways. First, this data provides a realistic estimate of the laboratory's confidence in any single result. Thus, if the instrument variability is ±1% and sample processing adds ±2%, each clinical result from that laboratory should fall within ±3%. Knowing this range allows the laboratory director to discuss CD4+ T-lymphocyte results objectively with practicing clinicians. Secondly, the assessment of laboratory variability sets a standard for lab performance that can be used to qualify technologists during their FCI training and to monitor proficiency within the lab.

The level of acceptable reproducibility between replicate CD34 enumerations has not been established. Replicate analyses of CD34 stains for the same specimens can be used to establish each laboratory's acceptable variation. A comparison of four replicate CD34 analyses for 23 different specimens is illustrated in Figure 6.5.

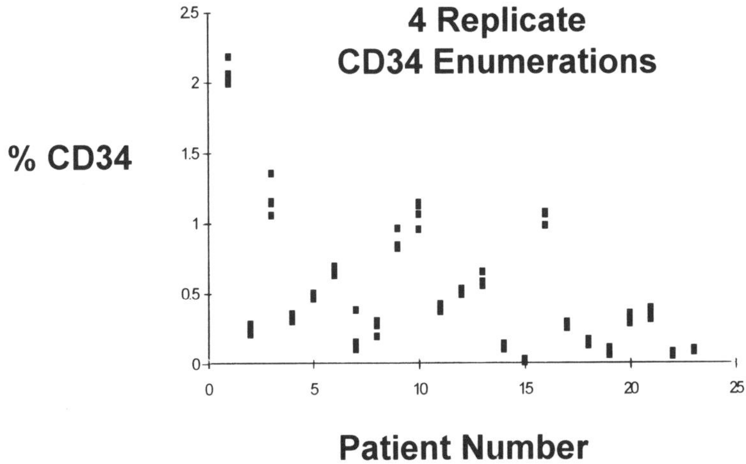

Fig. 6.5. Replicate CD34 analyses. Proper gating and the total number of positive cells counted can affect reproducibility, especially when positive cells are low in frequency. Procedural reproducibility of CD34 in four replicates from 23 patients was excellent in the study shown above.

Although the CD34+ cells in the majority of specimens were less than 1% of the total leukocytes, the agreement between the four determinations is excellent. The variation between results is limited not only by the proper gating procedures, but by the total number of positive cells counted in the sample. The lowest variability expected is approximately the square root of the total number of events counted. This means that for 100 CD34+ cells counted, the variation of this determination is approximately ±10 cells (10%).

Combining Hematology with Flow Cytometry

Most flow cytometers require hematology data in order to calculate absolute counts of subsets. The hematology analysis must include a complete white cell count and a differential.

To maintain consistent results, the hematology analysis should be performed in a standardized way. Quality assurance and quality control parameters must be monitored carefully. Variability between hematology instruments has been determined to be the greatest source of variability in flow cytometric immunophenotyping.

Proficiency testing reports indicate that this variability may be as much as 25%!

In an effort to remove the variability introduced by hematology analysis, some flow cytometers are now able to calculate absolute counts without any input from hematology. The results of this approach, if all of the flow cytometric quality assurance controls are satisfied, will be immunophenotyping results with less analytic variability.

BIOLOGIC VARIABILITY

There are a wide variety of biologic factors that can affect CD4 values. Age, sex, ethnic background, and diurnal variation have all been shown to have some effect on CD4 values.

Age

Age is of primary importance in evaluating CD4 levels in children versus those in adults and adolescents; the normal expected values of CD4+ lymphocytes are much higher in children. Studies in adults have also indicated some increases in CD4 with increasing decades of age, especially in women. A study by Reichert in 1991 reported a 1.1% increase in CD4 values per decade in both men and women (Fig. 6.6).

Gender and Race

Gender and ethnic background can influence reference ranges. Typically, reference ranges reflect heterogeneous populations and may be too broad unless they have been defined to be representative of the patient population being evaluated. For example, Asians generally have a larger number of natural killer cells, which can influence CD4 values if the precentages are not accounted for in gating and calculations.

There have been some studies that showed slight differences in CD4 values between races, but none of the differences were significant, as the studies did not include large enough populations. It is known, however, that CD4 epitopes recognized by certain monoclonal antibodies are not present in all Asians and blacks, so that CD4 cells may fail to be identified in these individuals.

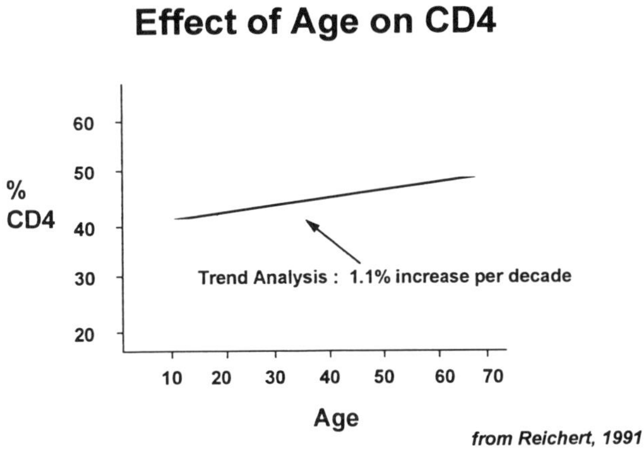

Fig. 6.6. Effect of age on CD4 values. A slight increase in CD4 values occurs in each decade of life, particularly in women. These increases are inconsequential compared to analytic variability of the measurement.

Gender differences have also been studied. Tollerud found that women had significantly more CD4 cells than men ($p = 0.01$). Bofill also reported higher CD4 percentages in women, but noted that most of the specimens drawn from women were drawn in the afternoon, while samples from the men studied were drawn in the morning. The differences shown in their study could thus be explained by diurnal variation alone.

In a more recent study of Northern European adults, Reichert found significant differences between men and women and calculated that CD4 values were 3.5% lower for men than for women.

Diurnal Variation

The greatest biologic influence may be caused by circadian rhythms, which have been reported in lymphocytes. Normal diurnal variation, reported by several investigators, may represent as much as 500-cell/μL difference in CD4 values between 10 p.m. and 8 a.m. in persons who are not infected with HIV (Fig. 6.7). Although HIV-seropositive individuals have a lesser diurnal variation in CD4 counts, values can increase by nearly 60 cells from morning to night according to a report by Malone in 1990.

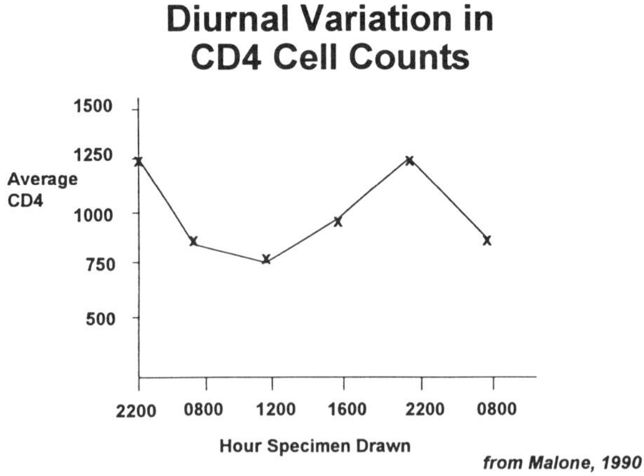

Fig. 6.7. Effect of diurnal variation on CD4 values. Circadian rhythm can affect lymphocytes and is of particular importance in quantitative immunophenotyping. In individuals who are not infected with HIV, CD4 values can change as much as 500 cells/μL from 10 p.m. to 8 a.m.

Physiologic Variables

What physiologic situations can affect flow cytometric lymphocyte quantitation? A wide variety of scenarios have been explored for CD4+ T lymphocytes, although the effect of these factors on CD34+ cells is not known. An extensive review of the literature was conducted by Owens, Stiehm, and Giorgi, from 1982 to the present, to identify reports of CD4 in natural or disease states and to critique the value of those reports in the context of today's assessment of quality assurance and quality control in CD4 immunophenotyping by flow cytometry. Data were reported for the following situations: infectious disease, immunologic disease, malnutrition, pregnancy, endocrine syndromes, radiation and therapeutics, and exposure to heavy metals. The results of this review are summarized below.

Infectious disease. As the host defense to infectious agents is the primary responsibility of the immune system, it is not surprising that many investigators have evaluated the effect of a variety of infectious agents on CD4+ T-lymphocyte levels. The influence of viral, fungal, bacterial, and parasitic infections on normal immune parameters has been reported.

Several groups evaluated the effect of the herpes-type viruses on CD4 levels; however all the groups measured CD4 cells after ficoll-hypaque separation. A large study by Gratama, using single-parameter flow cytometry, concluded that CD4 and CD8 were lower in all groups seropositive for Epstein-Barr virus, Herpes simplex virus or VZV, but none of the differences were statistically significant. Only the cytomegalovirus-seropositive group showed a significant increase in CD4 and CD8 cells in 334 patients ($p < 0.01$). Chronic fatigue syndrome, which may be caused by a herpes-type virus, was associated with a decrease in CD4 cells in a study of 35 patients.

Other viral diseases, such as measles, papillomavirus, and hepatitis B may cause a decrease in CD4 counts.

Bacterial meningitis may cause increases in the number of CD4+ cells, but other bacterial and fungal infections appear to cause decreased CD4 levels. Parasitic infections, such as *Cryptosporidium, Toxoplasma, Leishmania,* and *Strongyloides,* generally, but not always, cause decreased CD4 values.

Immunologic disease. It is of critical interest to evaluate T-cell subsets in patients with autoimmune disorders. Several types of autoimmune disease, allergy, asthma, and immunodeficiencies have been studied.

In general, autoimmune diseases are associated with lower CD4 levels, as reported for progressive multiple sclerosis, Bell's palsy, epilepsy, autism, systemic lupus erythematosus, Sjogren's syndrome, ataxia telangectasia, and rheumatoid arthritis. Some immunologic diseases, such as schizophrenia, IgA nephropathy, hereditary angioedema, and vitiligo, were reported to show significant increases in CD4 cells compared with controls.

Although several reports predicted that bronchial allergen exposure would cause a decrease in circulating CD4 cells and retention of these cells in the lungs, there is scant data to support this. In asthmatic patients, there are limited data that support a decrease in circulating CD4 cells after challenge.

Patients with common variable immunodeficiency may have smaller numbers of circulating CD4 lymphocytes, although the CD4 percentage appears to be normal.

Malnutrition. Nutritional status can affect the number and function of lymphocyte subsets. Both protein starvation and vitamin deficiencies may decrease T-cell subsets; however, there are no data to support any change in CD4 levels.

Pregnancy. The measurements obtained on lymphocyte subsets during pregnancy is affected not only by hormone stimulation, but also by stress and the increased volume of cell distribution. As a result, several investigators suggest that only percentages of subsets should be considered during pregnancy, not absolute numbers, which may be affected by the increased blood volume.

Many groups have reported on lymphocyte subsets during pregnancy. The first two trimesters show decreases in absolute numbers of CD4 cells, but not necessarily in their percentage. The numbers return to normal levels in the third trimester and postpartum.

There have actually been more studies of T-cell subsets in women with complicated pregnancies than in those with normal pregnancies. In HIV-infected pregnant women and pregnant women who smoke cigarettes, the normal decreases in CD4 during pregnancy were present, but smaller than normal.

Endocrine syndromes. Hematopoietic cells are known to be affected by normal hormonal fluctuations of the endocrine system. In addition to pregnancy, other conditions, such as diabetes, thyroid disorders, and Paget's disease of bone, have been studied for their influences on T-cell subsets.

In diabetics, the level of CD4 cells is increased, but levels may be lower than normal in prediabetes and new-onset diabetes. Patients with Paget's disease have been reported to have significantly lower absolute CD4 counts than sex- and age-matched controls. Individuals with thyroid disorders may present with either slight increases or slight decreases in CD4 levels.

Radiation and therapeutics. Radiation, chemotherapy, and immunostimulations can obviously affect cells of the immune system. Cancer patients, whether treated by chemotherapy or not, are known to show immunosuppression and have been characterized as having decreased CD4 levels. For CD34, however, immunostimulation is extremely important in the mobilization of stem cells for harvest. The concentration of CD34+ cells in the peripheral circulation varies with the type of induction therapy (Fig. 6.8).

Ultraviolet radiation is a common treatment for patients with many skin disorders. In general, such treatment causes a decrease in CD4 levels. Radiation therapy, using more penetrating radiation, can also cause a decrease in CD4 levels, sometimes lasting for long periods of time after treatment.

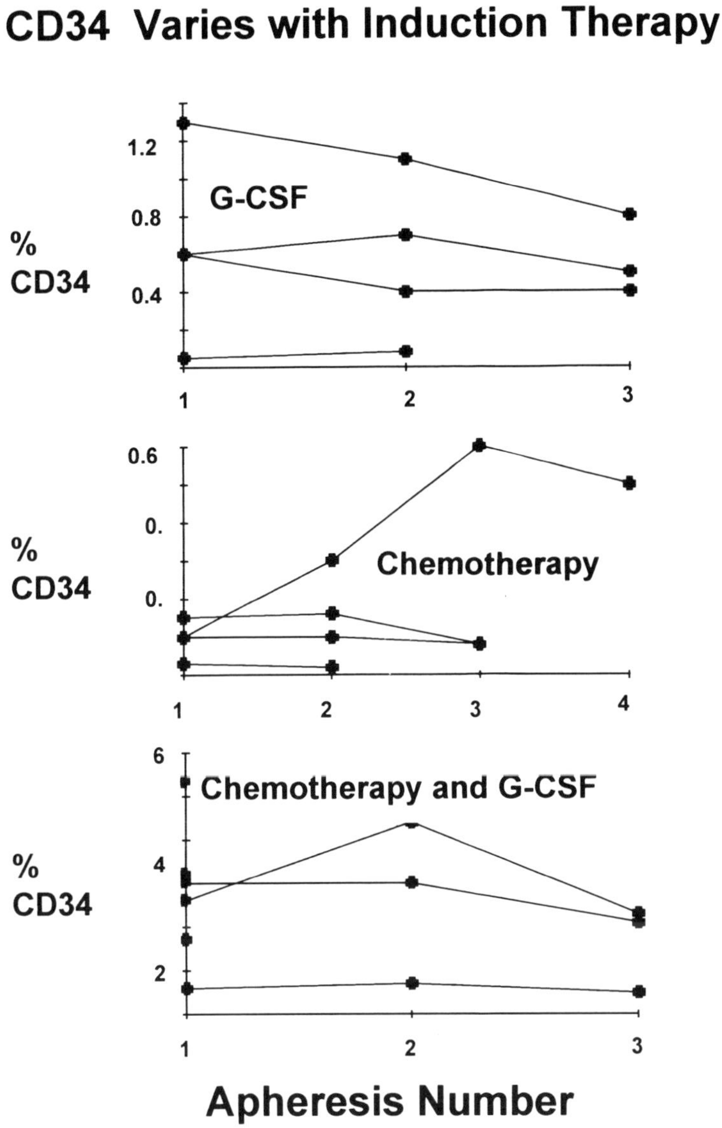

Fig. 6.8. Effect of induction therapy on the mobilization of CD34+ cells. The type of immunostimulation and the subsequent timing are both important factors in mobilizing the maximum number of PBSCs for harvest.

Heavy metals. Occupational exposure to mercury may cause an increase in the number of lymphocytes and in the absolute numbers of subsets, including CD4. In contrast, exposure to lead can cause a decrease in both percentages and absolute numbers of CD4 lymphocytes.

Lifestyle Variables

A second type of physiologic variability can be attributed to lifestyle factors. These external influences, such as smoking, drinking, physical exercise, and stress, are the result of personal choice and should be considered modifiable behavior.

Smoking. Smoking cigarettes has been found to cause an increase in WBC of all kinds. Many studies report an increase in CD4 levels in smokers, both men and women (Fig. 6.9).

In studies by Tollerud in 1989, Caucasians were reported to show this smoking-related increase in CD4, but a black population of smokers was determined to have significantly lower numbers of CD4 cells than nonsmokers.

Smoking during pregnancy also causes an increase in CD4 levels compared to those for nonsmoking pregnant women at the same stage of pregnancy.

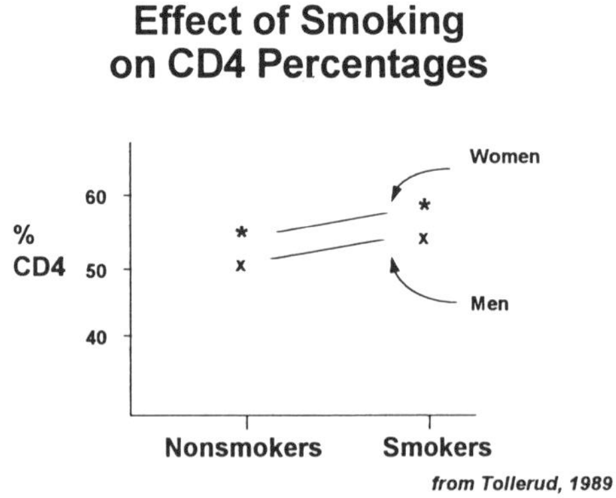

Fig. 6.9. Effect of smoking on CD4 values. Smoking cigarettes causes an increase in WBC and is of particular importance in quantitative immunophenotyping. CD4 levels can increase >3% in smokers. In the study shown here, women had higher CD4 values than men, both for smokers and nonsmokers.

HIV-seropositive smokers have increased CD4 levels, but the increase is blunted. In a study by Royce in 1990, cessation of smoking caused a decrease of 54 CD4 cells/mL.

Smoking marijuana also causes an increase in both the WBC and the CD4-lymphocyte subset.

Alcohol and drugs. While several studies using flow cytometric immunophenotyping have suggested that alcohol can cause an increase in numbers of WBC and lymphocytes, none have shown any significant changes in CD4 numbers.

No significant effects of amyl nitrite, methadone, or cocaine have been reported; however, active heroin users may have increased CD4 values.

Exercise and stress. Due to its effect on the pituitary-adrenal complex, various types of stress have long been suspected to cause fluctuations in lymphoid cell numbers. Physical exercise, thermal injury, and psychologic stress have all been studied.

At least 12 studies have concluded that WBC significantly increase in number immediately after strenuous exercise. The CD4 subset increases significantly in absolute numbers, but shows a decrease in percentage (Fig. 6.10). This difference in how results

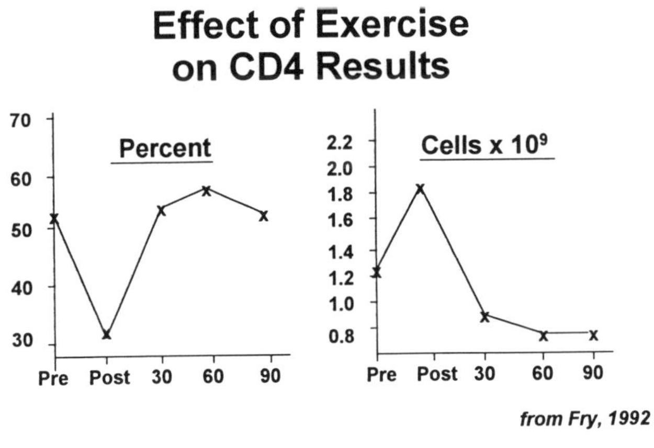

Fig. 6.10. Effect of physical exercise on CD4 levels. Immediately after exercise, the percent CD4 decreases and the absolute number of CD4 cells increases. By 30 minutes after exercise, the percent CD4 is normal or slightly elevated and the absolute CD4 counts have plunged to less than baseline levels.

are presented, (i.e., as absolute numbers or percentages) has created much confusion in the literature.

Thermal injury causes a decrease in CD4 percentage and may also decrease the CD4 absolute counts.

Psychologic stress, especially examination stress, may cause a significant decrease in percent CD4 without significant change in absolute CD4 levels.

REFERENCE RANGES

The whole point of assessing variability in CD4 values is to determine the analytic and biologic factors that can influence the accurate measurement of a cell population. These influences could affect a physician's interpretation of what is "normal" for a given patient.

The characterization of "normal ranges" for CD4 must be linked to the method of analysis; thus, there should be a "normal range" for CD4 cells by fluorescence microscopy, a "normal" range for CD4 cells by one-parameter FCI with gating by light scatter, a "normal" range for CD4+ T cells by two-parameter FCI with gating by light scatter, a "normal" range for CD4+ T cells by two-parameter FCI with gating by light scatter and fluorescence parameters, and an additional "normal" range for three-parameter analyses.

Good laboratory practice requires that reference ranges be revalidated whenever analysis techniques are changed significantly. In FCI, the techniques have changed faster than can be accomodated by appropriate validation studies. Furthermore, as clinical studies proliferate, it is not always clear which population is the appropriate "reference" for each patient.

"Normal" Reference Ranges

For a reference range to be statistically valid, it must be obtained from a significantly large population. For example, at least 250 individuals must be tested to obtain a reference range for CD4. It is best if these specimens are all drawn at the same time to avoid diurnal variation and that the time restrictions be stated along with the ranges.

The best way to state a reference range depends on the distribution of values in the study (Fig. 6.11). If the population follows a Gaussian distribution, then a range of two standard deviations is acceptable. If the population is not normally distributed, the data must be transformed before calculating a range of two standard deviations. Alternatively, the values at the 5th and 95th percentiles can also be used for a reference range.

Most published reference ranges for CD4 cells are based on a single-parameter definition of these cells, that is, lymphocytes are identified as CD4+ by monoclonal staining. In general, these ranges include all adolescents and adults, regardless of age, gender, or ethnic background.

If an individual patient cannot be compared to a perfectly matched reference range, by age, ethnicity, gender, or analytic methodology, what should a physician do to evaluate the significance of CD4 results?

First, the patient should be matched to the reference range most like the patient: adult, HIV-seronegative, North American, etc. Most of the published ranges are quite broad and include all but the highest and lowest values measured in apparently healthy individuals. If the patient's history includes other factors that may influence the CD4 result (e.g., smoking), the physician needs to estimate how much of an influence this factor would have on CD4 results.

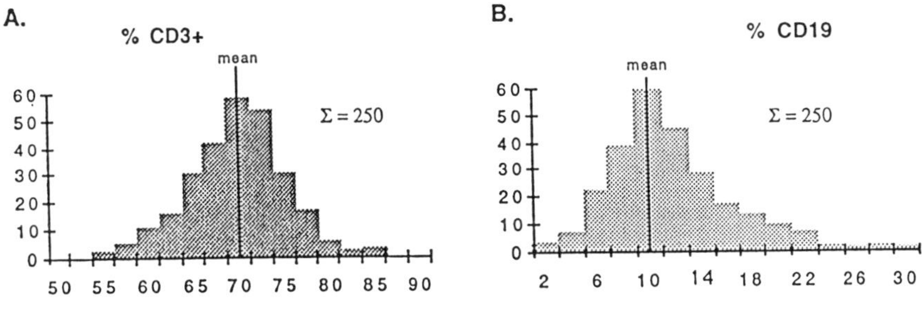

Fig. 6.11. Distribution of normal values. Some biologic parameters are normally distributed, like CD3, and can be described with a 2 SD range. Most biologic analytes, however, are not distributed over a Gaussian curve, like CD19, and data must be transformed before a 95% confidence range can be calculated.

Second, clinical decisions should not be based on a single laboratory test. A patient being followed for CD4 levels should have at least two determinations to substantiate a clinical finding of levels below 500 cells/μL or below 200 cells/μL. Using current flow cytometric techniques, the variability between replicates of the same specimen is quite low; however, the "within-person" variability on replicate specimens in a multi-site study has been reported to be 10.3–15.7% in 2787 HIV-seronegative homosexuals and 8.5% in a much smaller heterosexual control group.

In a retrospective analysis of the Multicenter AIDS Cohort Study, the T-cell subset results on 4954 homosexual men followed over 6 years were mathematically transformed and evaluated for analyte variability. The 95% confidence interval for the "true" CD4 value of a patient with a single CD4 measurement of 500 cells/μL was calculated to be 297–841 cells/μL and the interval for 200 cells/μL was 118–337 cells/μL. Hoover suggests that confirmatory retesting of these values reduced the probability of error to 4%.

Idiopathic CD4 Lymphocytopenia

A newly defined syndrome, idiopathic CD4 lymphocytopenia (ICL) is characterized by low levels of CD4+ T lymphocytes in the absence of HIV infection or therapeutic immunodeficiency. The definition of this syndrome relies on accurate measurements of CD4 cells, requiring two observations of CD4 $<$300 cells/μL. Individuals may present with a variety of viral, bacterial, fungal, and parasitic infections.

It has been suggested that the frequency of ICL is consistent with the number of "normal" individuals who would have $<$200 CD4 cells/μL in a Gaussian distribution of a large population. An alternative suggestion is that individuals who have ICL are simply those who develop late manifestations of congenital immune defects or perhaps late-onset primary T-lymphocyte deficiency. Since there have been no indications that ICL is caused by a transmissible agent, this extremely rare and clinically heterogeneous condition may simply occur in individuals who "normally" have a low number of circulating CD4 T lymphocytes.

CONCLUSIONS

Variability in quantitative flow cytometric immunophenotyping can directly affect patient management. All steps in the analytic process, from blood collection to the calculation of results, are potential sources of variability, which should be monitored as part of an effective quality assurance program. Biologic variability can also affect FCI results and include genetic, physiologic, and lifestyle factors. The interpretation of quantitative flow cytometric results may require clinicians to consider a variety of factors in addition to whether their patient falls within a "reference range."

FURTHER READINGS

Babcock GF, Taylor AF, Hynd BA, Sramkoski RM, Alexander JW. Flow cytometric analysis of lymphocyte subset phenotypes comparing normal children and adults. *Diagnostic and Clinical Immunology* 1987, 5:175–179.

Biggar RJ, Pahwa S, Minkoff H, Mendes H, Willoughby A, Landesman S, Goedert JJ. Immunosuppression in pregnant women infected with human immunodeficiency virus. *Am J Obstet Gynecol* 1989, 161:1239–1244.

Biselli R, Farrace S, D'Amelio R, Fattorossi A. Influence of stress on lymphocyte subset distribution—a flow cytometric study in young student pilots. *Aviat Space Environ Med* 1993, 64:116–120.

Centers for Disease Control. Guidelines for the performance of CD4+ T-cell determinations in persons with human immunodeficiency virus infection. *MMWR Morb Mort Wkly Rep* 1992, 41(RR-8):1–17.

Denny T, Yogev R, Gelman R, Skuza C, Oleske J, Chadwick E, Cheng S-C, Connor E. Lymphocyte subsets in healthy children during the first 5 years of life. *JAMA* 1992, 267:1484–1488.

Ekong T, Hill AM, Gompels M, Brown A, Pinching AJ. The effect of the temperature and duration of sample storage on the measurement of lymphocyte subpopulations from HIV-1-positive and control subjects. *J Immunol Meth* 1992, 151:217–225.

Fry RW, Morton AR, Crawford GPM, Keast D. Cell numbers and in vitro responses of leucocytes and lymphocyte subpopulations following maximal exercise and internal training sessions of different intensities. *Eur J Appl Physiol* 1992, 64:218–227.

Gelman R, Cheng SC, Kidd P, Waxdal M, Kagan J. Assessment of the effects of instrumentation, monoclonal antibody, and fluorochrome on flow cytometric immunophenotyping: a report based on 2 years of the NIAID DAIDS flow cytometry quality assessment program. *Clin Immunol Immunopathol* 1993, 66:150–162.

Giorgi JV, Cheng HL, Margolick JB, Bauer KD, Ferbas J, Waxdal M, Schmid I, Hultin LE, Jackson AL, Park L. Quality control in the flow cytometric measurement of T-lymphocyte subsets: the multicenter AIDS cohort study experience. The Multicenter AIDS Cohort Study Group. *Clin Immunol Immunopathol* 1990, 55:173–186.

Gratama JW, Naipal AM, Oosterveer MA, Stijnen T, Kluin-Nelemans HC, Ginsel LA, den Ottolander GJ, Hekker AC, D'Amaro J, van der Giessen M. Effects of herpes virus carrier status on peripheral T lymphocyte subsets. *Blood* 1987, 70:516–523.

Homberger HA, McCarthy R, Deodhar S. Assessment of interlaboratory variability in analytical cytology. Results of the College of American Pathologists Flow Cytometry Study. *Arch Pathol Lab Med* 1989, 113:667–672.

Hoover DR. Would confirmatory retesting of CD4+ cells to verify AIDS status be too expensive? [letter]. *Journal of Acquired Immune Deficiency Syndromes* 1993, 6:537–539.

Hughes MD, Stein DS, Gundacker HM, Valentine FT, Phair JP, Volberding PA. Within-subject variation in CD4 lymphocyte count in asymptomatic human immunodeficiency virus infection: implications for patient monitoring. *J Infect Dis* 1994, 169:28–36.

Koepke JA, Landay AL. Precision and accuracy of absolute lymphocyte counts. *Clin Immunol Immunopathol* 1989, 52:19–27.

Laurence J. T-cell subsets in health, infectious disease, and idiopathic CD4+ T lymphocytopenia. *Ann Internal Med* 1993, 119:55–62.

La Via MF, Hurtubise PE, Parker JW, Rubin RH, Burdash NM, Spencer-Green G, Bishop DA, Dziurzynski B. T-lymphocyte subset phenotypes: a multisite evaluation of normal subjects and patients with AIDS. *Diagnostic Immunology* 1985, 3:75–82.

Malone JL, Simms TE, Gray GC, Wagner KF, Burge JR, Burke DS. Sources of variability in repeated T-helper lymphocyte counts from human immunodeficiency virus type 1-infected patients: total lymphocyte count fluctuations and diurnal cycle are important. *Journal of Acquired Immune Deficiency Syndromes,* 1990, 3:144–151.

National Committee for Clinical Laboratory Standards. *Clinical Applications of Flow Cytometry, Quality Assurance and Immunophenotyping of Peripheral Blood Lymphocytes* (H42-T). National Committee for Clinical Laboratory Standards, Villanova, PA, 1992.

Nicholson JK, Jones BM, Cross GD, McDougal JS. Comparison of T and B cell analyses on fresh and aged blood. *J Immunol Methods* 1984, 73:29–40.

Owens MA, Stiehm ER, Giorgi JV. Lymphocyte measurements: In sickness and in health. Submitted for publication.

Paxton H, Kidd P, Landay A, Giorgi J, Flomenberg N, Walker E, Valentine F, Fahey J, Gelman R. Results of the flow cytometry ACTG quality control program: analysis and findings. *Clin Immunol Immunopathol* 1989, 52:68–84.

Reichert T, DeBruyere M, Deneys V, Totterman T, Lydyard P, Yuksel F,

Chapel H, Jewell D, Van Hove L, Linden J. Lymphocyte subset reference ranges in adult Caucasians. *Clin Immunol Immunopathol* 1991, 60:190–208.
Rickman WJ, Waxdal MJ, Monical C, Damato JD, and Burke DS. Department of Army lymphocyte immunophenotyping quality assurance program. *Clin Immunol Immunopathol* 1989, 52:85–95.
Robinson G, Morgan L, Evans M, McDermott S, Pereira S, Wansbrough-Jones M, Griffin G. Effect of type of haematology analyser on CD4 count [letter]. *Lancet* 1992, 340:485.
Royce RA, Winkelstein W Jr. HIV infection, cigarette smoking and CD4+ T-lymphocyte counts: preliminary results from the San Francisco Men's Health Study. *Aids* 1990, 4:327–333.
Signore A, Cugini P, Letizia C, Lucia P, Murano G, Pozzilli P. Study of the diurnal variation of human lymphocyte subsets. *J Clin Lab Immunol* 1985, 17:25–28.
Straus SE, Fritz S, Dale JK, Gould B, Strober W. Lymphocyte phenotype and function in the chronic fatigue syndrome. *J Clin Immunol* 1993, 13:30–40.
Tollerud DJ, Brown LM, Blattner WA, Mann DL, Pankiw-Trost L, Hoover RN. T cell subsets in healthy black smokers and nonsmokers. Evidence for ethnic group as an important response modifier. *Am Rev Respir Dis* 1991, 144(3 pt1):612–616.
Tollerud DJ, Clark JW, Brown LM, Neuland CY, Mann DL, Pankiw-Trost LK, Blattner WA, Hoover RN. The effects of cigarette smoking on T cell subsets. A population-based survey of healthy Caucasians. *Am Rev Respir Dis* 1989, 139:1446–1451.
Tollerud DJ, Clark JW, Brown LM, Neuland CY, Pankiw-Trost LK, Blattner WA, Hoover RN. The influence of age, race, and gender on peripheral blood mononuclear-cell subsets in healthy nonsmokers. *J Clin Immunol* 1989, 9:214–222.
Waxdal MJ, Monical MC, Fleisher TA, Marti GE. Inter-laboratory survey of lymphocyte immunophenotyping. *Pathology and Immunopathology Research* 1988, 7:345–356.

STUDY QUESTIONS

1. List three reasons why whole blood lysis methods are recommended for FCI.
2. Trace the evolution of immunophenotyping methodology from microscopy to three-color FCI.
3. What are the recommended anticoagulants for CD4+ T-lymphocyte testing?
4. What is the recommended anticoagulant for CD34 quantitation?

5. At what temperature should blood specimens be maintained for FCI?
6. Differentiate between "within-sample" reproducibility and "between-sample" reproducibility.
7. What is the variability in CD4 values between ethnic groups?
8. What is the variability between CD4 values in men and women?
9. What factor exerts the greatest influence on biologic variability in CD4 values?
10. Describe the influence of viral, bacterial, fungal, and parasitic infections on CD4 levels.
11. What effect does autoimmune disease have on CD4 values?
12. Explain why CD4 absolute values decrease during the first two trimesters of pregnancy while percent values of CD4 do not.
13. What is the effect of smoking on lymphocyte counts?
14. What is the effect of smoking on CD4 counts? In pregnant women? In HIV-seropositive smokers?
15. What happens to CD4 values after strenuous exercise?
16. Does psychologic stress have any effect on CD4 levels?
17. Why is it difficult to find published reference ranges for specific ages and ethnic groups?
18. How should a reference range be validated?
19. Estimate the variability of a CD4 value of approximately 200 cell/μL.
20. Explain why idiopathic CD4 lymphocytopenia could be a "normal" situation.

VII

Safety and Specimen Acquisition, Handling, and Processing

INTRODUCTION

Safety in the handling of clinical specimens is a requirement for good laboratory practice. Safe handling procedures will provide some standardization for flow cytometric immunophenotyping as safety restrictions eliminate certain practices. The correct handling of specimens for FCI also requires adherence to procedures designed to standardize specimens for optimal stability through restrictions on the anticoagulant used, age of specimen, and storage temperature.

DEVELOPING SAFETY GUIDELINES

Guidelines for safety are required because many hazards exist in the laboratory environment. Making the laboratory safe is everyone's responsibility. There are known hazards related to the processing of any human materials; thus it is important to have standard procedures to protect against these hazards.

Safety is based on logic and common sense. There may be situations in which employees or visitors do not know what hazards are present in the laboratory. At such times, everyone should behave as though a hazardous situation exists. An effective, operating safety committee in each laboratory is essential to identify risks, to develop safe working policies, to train staff in approved practices, and to ensure that safety policies are followed. Everyone must be committed to ensuring a safe work environment and be empowered to challenge others on unsafe practices.

Phlebotomy Safety

Phlebotomy safety must be included as part of a comprehensive safety program and must follow the CDC's universal precautions. Proper training is required to perform phlebotomy correctly. It is essential to protect the patient from potential infection and the phlebotomist from exposure to biologic hazards. It is also essential to protect cleanup personnel from biologic waste, including needles and other sharp objects that could cause cuts or puncture wounds. In addition, it is important to protect the specimen to ensure that the best results are obtained.

Guidelines issued by the National Committee for Clinical Laboratory Standards (NCCLS) for drawing blood must be made available for review and comment by phlebotomists. The most important safety issue is the care and handling of sharp objects that may cause puncture wounds. It is important not to recap needles and to dispose of sharp objects in puncture-proof containers designed for that purpose.

Hazards Related to Flow Cytometry

In addition to routine safety practices that operate in any workplace, there are safety issues which are unique to the flow cytometry laboratory.

Laser hazards. Laser light is a potential hazard found in a flow cytometry laboratory. It is important to follow the manufacturer's recommendations for safe operation of the instruments and to avoid placing body parts or reflective objects in the laser beam.

Biologic hazards. Biologic hazards are frequently present in the flow cytometry laboratory. Because the clinical laboratory processes human biologic material, there are potential infectious agents that may be present in specimens. The risks and consequences have increased due to the possibility of HIV-infected blood and the unusual pathogens that may be present in the blood of AIDS patients.

Although many human diseases can be passed from specimen to laboratory worker, two of the most serious are hepatitis and HIV.

A laboratory worker can be exposed in a number of ways. The risks include direct puncture wounds by sharp objects, exposure via aerosol, and indirect exposure by ingestion (eating or drinking) and by hand-to-mouth or hand-to-wound contact.

Both hepatitis and HIV, as well as other biologic agents, can be inactivated by diluted household bleach, commerical viricidal agents, or paraformaldehyde. Paraformaldehyde should be added to all flow cytometry samples at a final concentration of 1% in order to inactivate hepatitis and HIV viruses. For safety, all cells should remain in the paraformaldehyde solution for flow analysis.

Chemical hazards. Most flow cytometry laboratories use chemicals that present multiple hazards to the laboratory worker. Solvents must be stored properly to minimize potential fire hazards. Acids and bases must be stored properly and labeled correctly to minimize potential exposure hazards.

Poisons used to inhibit bacterial growth in reagents and test samples are also commonly found in the flow cytometry laboratory. The most common poisons are sodium azide, which is included in most commerical antibody preparations, and paraformaldehyde or formaldehyde, which is used in FCI cell fixation.

Sodium azide poses an additional danger in that it forms explosive compounds when allowed to build up in lead pipe plumbing. Therefore, it is important to use extreme care in the disposal of monoclonal antibody reagents containing sodium azide. Solutions containing sodium azide must be flushed down the drain using copious amounts of water.

Other common chemical hazards encountered in a flow cytometry laboratory are nucleic acid stains, which are potential carcinogens. While these compounds are generally used in dilute solutions, extra care must be taken in preparing stock solutions from powder or concentrate.

LABORATORY SAFETY

A safe laboratory requires clearly defined procedures and laboratory discipline. All steps in the handling of biohazardous materials must follow universal precautions. Always wear proper laboratory attire when operating a flow cytometer. This includes latex gloves and a laboratory coat. Some labs also require safety glasses in the FCI area.

Install plastic covers on computer keyboards to protect them from spills and for easy cleanup and disinfection. Designate certain computers as "clean" where no gloves or potentially contaminated materials are allowed. Permit absolutely no eating, drinking, or application of cosmetics within a laboratory.

Prepare and make readily available an emergency spill cart for handling accidents that may occur in the laboratory. A suitably equipped, portable cart should be available for major spills. Put disinfectants within easy reach in the hood and on the laboratory bench and train personnel in proper decontamination procedures. Provide clearly written procedures for handling all types of potential spills.

Laboratory Attire

Universal laboratory precautions include specification of laboratory attire, also called "personal protective equipment."

1. Wear laboratory coats and remove them when leaving the laboratory.
2. Do not wear open-toed shoes in the laboratory.
3. Make proper use of disposable latex gloves. Always wear disposable gloves when handling specimens. Remove gloves before entering "clean" areas and dispose of them immediately in a biohazard container.
4. Wear masks and protective eyewear if any work is to be performed that presents a risk of mucous membrane exposure to blood or body fluids.
5. Wash hands frequently and immediately after completing laboratory activities.
6. Provide "clean" areas where laboratory attire is forbidden.

Biosafety Hood

Universal laboratory precautions define procedures for working in a biosafety hood. Clinical specimens and test sample tubes must be processed in a Class 1 or Class 2 biosafety hood. Control or minimize aerosols. Although the most hazardous step in specimen processing is opening the vacuum container, aerosols can be formed at any stage of sample processing.

No pipetting by mouth is permitted. Have mechanical pipetting devices at all pipetting stations. Discard pipette tips and disposable pipettes in a biohazard container. Use care when pipetting to minimize drips from pipettes. Disinfect drips immediately when they occur.

Check vacuum traps to be sure they contain the proper amount of disinfectant. Assign clear responsibility for cleaning vacuum traps and check to see that this is being done routinely. Always dispose of sharp objects in puncture-resistant and leak-proof biohazard containers.

Centrifuge Safety

Universal precautions for the laboratory extend to centrifuge safety.

1. Use plastic tubes for all samples.
2. Recap tubes between individual steps in the procedures to reduce the potential for spills and aerosols. This is especially important when mixing or vortexing a sample.
3. Use sealed centrifuge cups to reduce potential aerosols or possible spills.
4. Clearly define procedures for decontaminating centrifuges following a spill, and for regular cleaning and maintenance.

Waste Disposal

All blood specimens and sample tubes must be treated as potentially biohazardous and disposed of properly. In addition, all materials that come in contact with specimens must also be treated as biohazardous waste. It is important to protect against puncture wounds by placing any sharp objects in appropriate puncture-resistant and leak-proof containers.

The samples for flow cytometry contain paraformaldehyde and should be treated as biohazardous chemical waste. Most specimen samples should be decontaminated, preferably by autoclaving or by incinerating. Chemical decontamination with 0.5% sodium hypochlorite (10% household bleach) is acceptable but necessitates incineration since chlorine-containing material can damage autoclaves.

Disinfecting the Flow Cytometer

The flow cytometer should be periodically disinfected by aspirating a 1 : 10 dilution of household bleach through the flow cytometer to decontaminate the system after operation. Bleach must be run for at least 5 minutes at the end of each day. It is important to rinse out the bleach with saline for at least 15 minutes as bleach left in the fluidics can cause rapid deterioration of the lines. Concentrated bleach may be used to remove plugs and buildup of biologic materials. All external surfaces should be wiped down with diluted bleach, or a phenol or alcohol disinfectant, at the end of every day.

Waste from the flow cytometer must be decontaminated by adding household bleach to the waste fluid container to give a final concentration in a full waste container of 10% bleach. The bleach will disinfect the waste fuilds as they collect. In addition, stream-in-air instruments must be operated as recommended by the manufacturers for reduction of aerosols.

SPECIMEN COLLECTION AND TRANSPORT

Whole blood specimens are required for quantitative flow cytometric immunophenotyping (Fig. 7.1). The quality control of specimen processing begins with the blood collection itself.

Assumption: It is assumed that variations introduced during sample handling, processing, and analysis do not affect the results.

Universal precautions must be followed during all phases of specimen collection. Appropriate anticoagulants must be selected for both the hematology and the flow cytometry laboratories. Specimens should be drawn into EDTA, heparin, or acid citrate dextrose (ACD) depending on the anticipated elapsed time for collection until specimen processing in the testing laboratory, or on the specific specimen requirements.

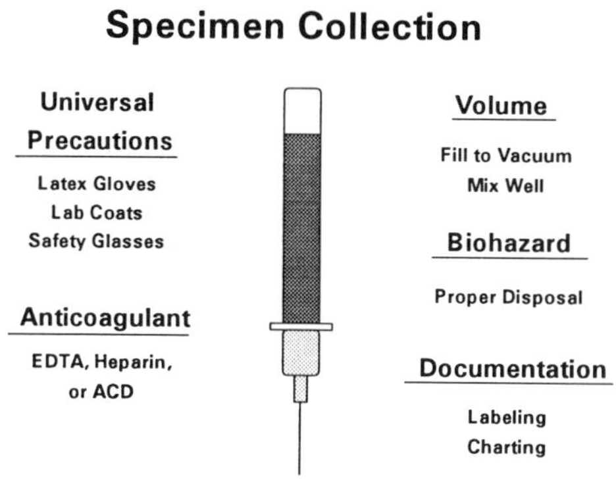

Fig. 7.1. Quality assurance and safety in specimen collection. All specimens for flow cytometric immunophenotyping must be collected in the proper anticoagulant, following universal precautions. Disposal must follow biohazard procedures. All documentation must be complete.

It is necessary to fill collection tubes completely, so that no vacuum remains in the tubes. Proper mixing of anticoagulant and blood is necessary to preserve the specimen. The phlebotomist should verify with the patient that all of the appropriate demographic information is correct. All forms must be filled out correctly and the specimen tubes must be labeled properly, immediately after the blood is drawn. The date and time of specimen collection should be recorded in the patient chart, on the test request form, and on the specimen tube label. Needles and other sharp materials must be discarded in a puncture-resistant biohazard container.

Anticoagulants

Two different testing procedures are required for obtaining quantitative CD4+ T-lymphocyte subset counts: FCI and hematology analysis. Anticoagulant requiremens for the two procedures may be different.

Specimen collection must be performed properly and blood must be collected into the appropriate anticoagulant for each of the procedures. The anticoagulant selected may depend upon the time period between specimen drawing and sample analysis (Fig. 7.2). Most automated hematology analyzers require the use of EDTA. Specimens to be run on a hematology analyzer should be processed

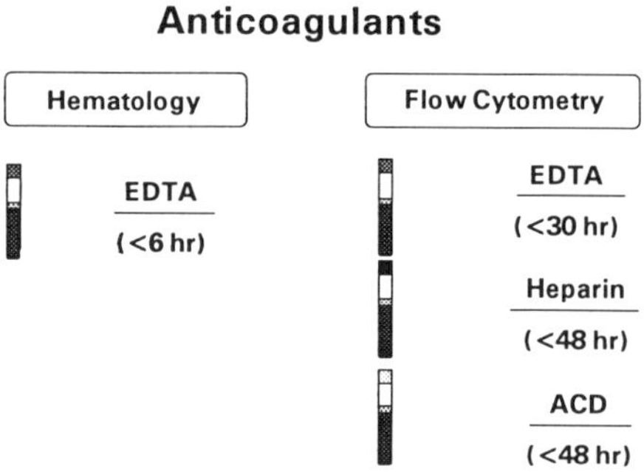

Fig. 7.2. Anticoagulants for quantitative CD4 immunophenotyping. The hematology specimen must be drawn in EDTA. The paired specimen for immunophenotyping may be drawn in EDTA (if it will be processed in <30 hours) or in heparin or ACD for stability up to 48 hours.

within 6 hours of collection. The instructions for some instruments claim that specimens are stable up to 24 hours after collection, but each laboratory should validate manufacturers' claims before adjusting specimen requirements. If the specimen is processed within 30 hours of venipuncture, EDTA can also be used for FCI. Immunophenotyping light scatter characteristics are poorly preserved with EDTA specimens that are held more than 24 hours and are unacceptable for those held beyond 30 hours.

Only one specimen tube containing EDTA is required when hematology and FCI are both performed in the same laboratory and when FCI can be done within 30 hours of blood collection. When hematology and FCI testing are performed in two different laboratories, one specimen tube is collected into EDTA and another tube is collected into either EDTA, heparin, or ACD anticoagulant, depending on the time delay before FCI testing.

Blood specimens that require longer than 1 day to reach the flow cytometry laboratory should be drawn into either heparin or ACD anticoagulant. Either of these anticoagulants preserve the cellular characteristics better over time than does EDTA. Specimens older than 48 hours after blood draw are unacceptable for flow cytometry and should not be processed, no matter what anticoagulant was used.

Blood collection tubes should be filled completely so that no vacuum remains. Eliminating excess vacuum in the collection tube minimizes the potential for an aerosol hazard when the tube is opened in the FCI laboratory. Completely filling the specimen collection tube also ensures that the proper amount of anticoagulant is used for maintaining specimen integrity. Gently inverting or shaking the blood collection tubes can facilitate mixing of blood and anticoagulant to prevent clotting. Mixing also ensures that all the blood in the tube is exposed to the same concentration of anticoagulant.

Pediatric specimens should be collected into pediatric-sized tubes. It should be noted that only a small volume of whole blood is required for immunophenotyping, so pediatric tubes are adequate for collecting blood specimens from adults.

Specimen Labeling

Proper labeling of the specimen is an important component of a quality control program. The specimen should be identified using the patient's name and/or a unique patient identifier to maintain patient confidentiality.

It is important, for several reasons, to include date and time of collection on the specimen label. Specimen age is important for sample processing, and laboratories should reject specimens that are more than 48 hours old. It is also important to note the time of collection, since there are significant diurnal fluctuations in lymphocyte subpopulations. When monitoring individual patients over time, specimens should be collected at the same time of day in order to minimize this source of variability.

The test request form must also be properly completed. Several biologic factors affect lymphocyte subset results. Age is important, especially in differentiating pediatric from adult specimens. Sex must be noted since small variations in lymphocyte subsets occur between sexes. Race should be indicated when possible, since there may be significant genetic differences in the reference ranges used for different races.

Any medications that the patient is taking should be included on the request form, since these may affect lymphocyte subset analysis. Drugs such as steroids, zidovudine (AZT), 2′,3′-dideoxyinosine (ddI), and other medications can significantly influence the proportion of lymphocytes identified by FCI. Disease conditions may also affect the results and should be noted on the form. Some examples include HIV seropositivity, other immunodeficiencies, chemotherapy, or organ transplantation.

Specimen Transport

Transporting specimens requires protecting both the specimen and any persons who may come in contact with it. The specimen must be protected from temperature extremes. Transport within a facility requires that specimen tubes be properly labeled and then placed in a zip-lock plastic bag or other leak-proof container. The plastic bag will contain the blood if the specimen tube is dropped or accidentally opens.

Transport or shipment of specimens to another site for testing requires adherence to federal, state, and local regulations as appropriate. Companies that handle shipments may have more stringent, specific requirements for packaging of diagnostic specimens. Therefore, the packaging should also conform to the requirements of the carrrier.

The Centers for Disease Control and Prevention recommend packaging biohazardous specimens in several layers of protection. Properly labeled specimen tubes should be placed in leak-proof containers such as sealed plastic bags. A number of patient specimen tubes may be shipped at the same time, each in a separate plastic bag. By using separate sealed plastic bags, any breakage should be contained and should not contaminate the other specimens. The plastic bags containing the specimen tubes are gathered together along with the appropriate test request forms and placed in a primary shipping container containing ample absorbent materials to soak up any blood resulting from leaks or broken tubes during shipment (Fig. 7.3).

The shipping container should have a rigid structure and be able to withstand normal shipping conditions. The inner shipping container(s) should be placed inside a secondary container containing cushioning materials. Cool packs can be placed in the outer shipping container to maintain ambient temperatures in the specimen container if extremely hot temperatures are anticipated.

All specimens should be maintained at a room temperature of 18°C to 22°C and shipped in insulated containers to minimize expo-

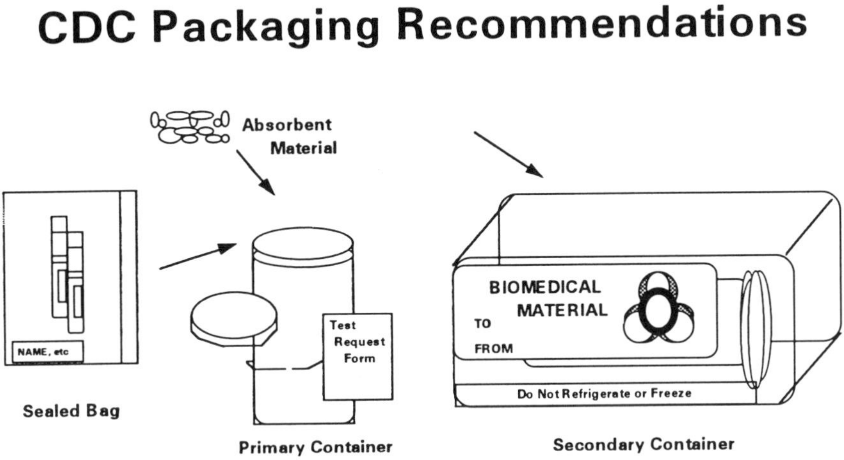

Fig. 7.3. Quality assurance in specimen packaging. The CDC recommends packaging biohazardous materials in several layers of protection. Blood specimens should be placed in zip-lock bags, inside a rigid primary transport container, which is inside a rigid secondary container. The specimen and all persons who handle the specimen must be protected.

sure to extremely high (>37°C) or low temperatures (<10°C) during transit to the testing laboratory. The exterior of all packages containing specimens for immunophenotyping must be labeled with biohazard warnings consistent with state and federal regulations and must indicate "Do Not Refrigerate or Freeze."

The outer package should also be labeled with the recipient's complete address and the telephone number of the contact person in the laboratory, especially if the delivery is to be made to a shipping or receiving department and not directly to the laboratory. Persons responsible for receiving shipments should be instructed to watch for the arrival of specimens and expedite delivery to the flow cytometry laboratory.

Specimen Inspection

Upon arrival in the laboratory, the specimen should be unpacked and inspected. Specimens should be logged into a tracking system immediately upon arrival at the laboratory. In addition, the apparent temperature of the specimen should be noted (hot or cold to the touch) and recorded in the appropriate log. Light scatter patterns will generally identify a specimen that has been damaged by extremely hot or cold temperatures during shipping or storage and may confirm initial indications of inappropriate storage temperature. Maintain all specimens at room temperature.

If a tube is broken or cracked, the specimen must be rejected and treated as biohazardous waste. A second specimen should then be requested immediately. If the blood is hemolyzed or frozen, or if clots are visible, the specimen should be rejected and another requested. No specimen more than 48 hours old should be processed for flow cytometric immunophenotyping.

SAMPLE PROCESSING

Patient specimens received by the laboratory are accessioned into the laboratory data system and transferred to the proper testing areas. Proper subset analysis generally requires that specimens be processed by the two different technologies of hematology and flow cytometry, and results from both are combined to determine the absolute cell counts.

Hematology Analysis

Hematology results are obtained from an EDTA specimen collected within the previous 6 hours. The hematology parameters that are most important for lymphocyte subset analysis are the white blood count (WBC), in total cells per microliter, and the automated differential count for percentage of lymphocytes in the specimen. Analysis of red cells or platelets is not considered in lymphocyte subset quantitation. Between 10,000 and 30,000 white blood cells should be analyzed to obtain an accurate WBC and differential.

If the sample is rejected or flagged based on either the WBC or differential, a manual differential of at least 400 cells should be performed and used in the calculations. However, if the error message from the automated hematology analyzer does not involve the lymphocyte population, the automated lymphocyte differential can be used. The results from the hematology analysis are reported to the flow cytometry laboratory for calculating absolute cell counts.

Assumption: It is assumed that the proportion of leukocytes or lymphocytes identified by either a flow cytometer or a hematology analyzer is identical.

Automated hematology analyzers and flow cytometry systems do not identify lymphocytes identically. A flow cytometer identifies lymphocytes based on light scattering, and lymphocyte identity is verified by CD45 and CD14 fluorescence staining. Flow cytometers also can utilize CD45 fluorescence and forward angle light scattering to identify the leukocytes.

Hematology analyzers, on the other hand, use a variety of techniques, depending on the instrument. Some instruments use electronic volume sensing, others use light scattering, while leukocyte stains and lysing protocols are used in others. As instruments define a lymphocyte somewhat differently, they may not give the same differential analysis for the same specimen.

Hematology analyzers have significant variability when analyzing specimens that are more than 6 hours old. When EDTA is used as the anticoagulant, neutrophils begin to die or degranulate after 6 hours. This causes an increased variability in both the WBC and differential count, so that the absolute lymphocyte count, a product of these measurements, must also be affected by viability.

Flow Cytometric Immunophenotyping

The FCI results are combined with the hematology results. Blood preserved with EDTA, heparin, or ACD is sent to the FCI laboratory for analysis. Aliquots of blood are placed in test tubes and incubated with fluorescent-conjugated monoclonal antibodies specific for known cell-surface antigens. After incubation, the erythrocytes are lysed and the remaining white cells are washed, concentrated, fixed in paraformaldehyde, and analyzed on the flow cytometer (Fig. 7.4).

Erythrocyte lysis. Erythrocyte lysis is an important component in specimen processing. Red blood cells (RBC) outnumber white blood cells (WBC) 1000 : 1 in normal blood. Therefore, in the immunophenotyping procedure, white blood cells could be considered rare events if RBC are not removed. Accurate lymphocyte analysis requires the removal of both the red cells and platelets.

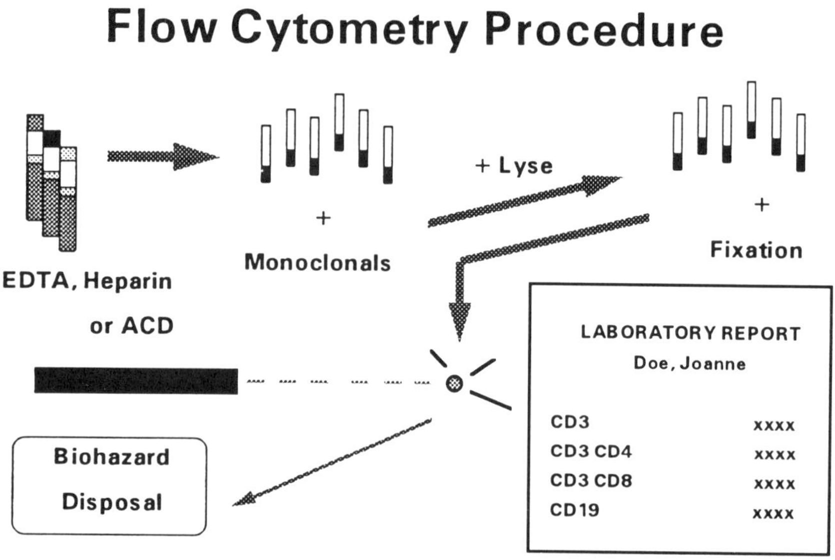

Fig. 7.4. Flow cytometric procedures. Aliquots from a patient's blood specimen are placed in tubes and stained with fluorochrome-labeled monoclonal antibodies, the red cells are lysed, and the remaining WBC fixed with paraformaldehyde. Cells are evaluated by flow cytometry and biohazardous waste disposed appropriately.

Density gradients. A density gradient is one way of removing erythrocytes and platelets; however, it is not recommended for use in immunophenotyping. Variable recovery of lymphocytes is observed using density gradients such as ficoll-hypaque, or percoll. The preferred method for processing cells for immunofluorescence is to stain the sample first, then lyse the erythrocytes, and finally fix the white cells. The platelets are removed during wash steps using low-force centrifugation (300–400 *g*).

Ammonium chloride. Several chemical procedures have been used to remove erythrocytes from blood samples. Ammonium chloride (NH_4Cl) reacts with carbon anhydrase present in erythrocytes to cause cell lysis. For some specimens, complete erythrocyte lysis may be difficult to attain using this procedure. In addition, certain medications can affect the sensitivity of erythrocytes to lysis. This procedure will not remove all nucleated red cells.

Commercial lytic agents. Several lysis procedures are commercially available, based on combinations of acid and enzymes. Some protocols follow the stain, lyse, wash, and fix procedure. Others use a procedure of stain, lyse, fix, and then wash. Some commerical lytic agents require a separate centrifugation step with a preferential loss of lymphocytes relative to other leukocytes. Timing of reagent addition for commerical lysis protocols is critical, so manufactuers' recommendations must be followed exactly. Changes made to lytic protocols must be validated within each laboratory. Prior to making any procedural changes, it is also important to assess the impact that switching from one lytic protocol to another has on final results.

Variability caused by lytic protocol. Variability in results can be introduced by the lytic procedure. Certain specimen processing procedures have a selected loss of lymphocytes relative to other cell types. The sensitivity of lymphocytes to lytic agents is different than that of erythrocytes and other leukocytes. There are also differences in sedimentation rates between lymphocytes and other leukocytes and these can result in differential recovery of lymphocytes after centrifugation. Manufacturers' recommended procedures should be followed exactly to minimize this variation in the results.

Light-scattering changes. Lymphocyte subsets can also undergo variable loss during different erythrocyte lysis procedures, and lysis protocols can also change the light-scattering properties of the samples. Therefore, specimen processing procedures should be standardized within a laboratory to minimize this source of variability in specimen processing.

In selecting a sample processing procedure, a laboratory may wish to compare two or more procedures to assess effects on the specimen. This can be done by processing aliquots of the same specimen in different ways and comparing the results. If changing the specimen processing procedure biases the results, only a small number of samples will be needed to show any differences.

Sample fixation. Sample fixation must be performed both to preserve biologic characteristics and to reduce biohazards. The fixative of choice for preserving immunofluorescence characteristics is a commerically prepared solution of paraformaldehyde or formaldehyde (electron microscope grade). A 1–2% paraformaldehyde solution should be made up each week from a commercially available concentrated liquid solution. Preparing fixative solutions from formaldehyde powders is not recommended because of dangers from inhalation of powder and toxic fumes. Phosphate buffered saline is used as the diluent and the pH should be adjusted to 7.2–7.4. If the fixative is not buffered it is possible to quench fluorescein fluorescence.

Samples should be capped and stored in the dark in a refrigerator (2–8°C) after the paraformaldehyde has been added. The 1% paraformaldehyde preserves the light scatter and fluorescence characteristics required for FCI. In addition, 1% paraformaldehyde inactiviates free HIV and hepatitis viruses within 15 minutes. It has not been established how long it takes to inactivate cell-associated virus, therefore, all samples should remain in fixative throughout the analysis on the flow cytometer.

After fixation, samples should be analyzed within 24 hours. There is a slow increase in autofluorescence observed in samples stored longer than 48 hours. This increase in autofluorescence may significantly affect the resolution of continuous and negative populations from one another. Adding the fixative while the cells are exposed to antibodies can increase nonspecific staining from the cross-linking of antibodies to cells.

CONCLUSIONS

All clinical specimens must be considered potentially biohazardous and the laboratory must establish safety procedures for handling specimens. The proper collection, transport, and processing of blood specimens for FCI must follow approved guidelines to ensure the stability of the specimen and the safety of all personnel.

FURTHER READINGS

Ashmore RM, Shopp GM, Edwards BS. Lymphocyte subset analysis by flow cytometry. Comparison of three different staining techniques and effects of blood storage. *J Immunol Methods* 1989, 118:209–215.

Calam R, Benoit S, Foreback C. *Procedures for the Handling and Processing of Blood Specimens* (H18-A). The National Committee for Clinical Laboratory Standards, Villanova, PA, 1990, p 34.

Carter PH, Resto-Ruiz S, Washington GC, Ethridge S, Palini A, Vogt R, Waxdal M, Fleisher T, Noguchi PD, Marti GE. Flow cytometric analysis of whole blood lysis, three anticoagulants, and five cell preparations. *Cytometry* 1992, 13:68–74.

Centers for Disease Control. Update: universal precautions for prevention of transmission of human immunodeficiency virus, hepatitis B virus, and other bloodborne pathogens in health-care settings. *MMWR Morb Mortal Wkly Rep* 1988, 37:377–382, 387–388.

Centers for Disease Control. Guidelines for the performance of CD4+ T-cell determinations in persons with human immunodeficiency virus infection. *MMWR Morb Mortal Wkly Rep* 1992, 41(RR-8):1–17.

Centers for Disease Control. Public Health Service statement on management of occupational exposure to human immunodeficiency virus, including considerations regarding zidovudine postexposure use. *MMWR Morb Mortal Wkly Rep* 1990, 39(RR-1).

Ekong T, Hill Am, Gompels M, Brown A, Pinching AJ. The effect of the temperature and duration of sample storage on the measurement of lymphocyte subpopulations from HIV-1-positive and control subjects. *J Immunol Methods* 1992, 151:217–225.

Koepke JA and Landay AL. Precision and accuracy of absolute lymphocyte counts. *Clin Immunol Immunopathol* 1989, 52:19–27.

Lal R, Edison L, Chused T. Fixation and long-term storage of human lymphocytes for surface marker analysis by flow cytometry. *Cytometry* 1988, 9:213–219.

Lifson J, Sasaki D, Engleman E. Utility of formaldehyde fixation for flow cytometry and inactivation of the AIDS-associated retrovirus. *J Immunol Methods* 1986, 86:143–149.

National Committee for Clinical Laboratory Standards. *Protection of Laboratory Workers from Infectious Disease Transmitted by Blood and Tissue* (M29-P). National Committee for Clinical Laboratory Standards, Villanova, PA, 1987.

National Committee for Clinical Laboratory Standards. *Procedures for the Domestic Handling and Transport of Diagnostic Specimens and Etiologic Agents* (H5-A2). National Committee for Clinical Laboratory Standards, Villanova, PA, 1985.

National Committee for Clinical Laboratory Standards. *Procedures for the Collection of Diagnostic Blood Specimens by Venipuncture* (H3-A3). 3rd ed. National Committee for Clinical Laboratory Standards, Villanova, PA, 1991.

National Committee for Clinical Laboratory Standards. *Protection of Laboratory Workers from Instrument Biohazards* (I17-P). National Committee for Clinical Laboratory Standards, Villanova, PA, 1991.

National Committee for Clinical Laboratory Standards. *Clinical Applications of Flow Cytometry, Quality Assurance and Immunophenotyping of Peripheral Blood Lymphocytes* (H42-T). National Committee for Clinical Laboratory Standards, Villanova, PA, 1992.

Nicholson JK, Jones BM, Cross GD, McDougal JS. Comparison of T and B cell analyses on fresh and aged blood. *J Immunol Methods* 1984, 73:29–40.

Occupational Safety and Health Administration (OSHA). Blood borne pathogens, (29 CFR Part 1910.1030). *Federal Register*, December 6, 1991, 56.

Occupational Safety and Health Administration ((OHSA). Occupational exposure to hazardous chemicals in laboratories (29 CFR part 1910,145). *Federal Register*, January 31, 1990, 55.

Robinson G, Morgan L, Evans M, McDermott S, Pereira S, Wansbrough-Jones M, Griffin G. Effect of type of haematology analyser on CD4 count [letter]. *Lancet* 1992, 340:485.

Thornthwaite, JT, Rosenthal PK, Vazquez DA, Seckinger D. The effects of anticoagulant and temperature on the measurements of helper and suppressor cells. *Diagnostic Immunology* 1984, 2:167–174.

STUDY QUESTIONS

1. Describe the precautions that should be taken to prevent janitors and clerical workers from being exposed to laboratory hazards.
2. Explain the importance of standard operating procedures to protect against known and unknown laboratory hazards.
3. List the three different categories of safety issues unique to the flow cytometry laboratory and give one example of a specific safety concern for each.
4. List two items of personal protective equipment that must be worn in the flow cytometry laboratory.
5. List three items that are important for using a centrifuge safely.
6. Give two reasons for installing a plastic cover on the computer keyboard.
7. Describe when and where laboratory attire, including latex gloves and lab coats, should be worn or not worn.
8. Describe the uses for a well-equipped emergency spill cart.
9. Describe how specimens should be prepared for in-house transport.
10. Describe how to package a blood specimen for off-site transport.
11. List the items that should be included on the specimen package label.

12. Explain the proper procedures for disinfecting the flow cytometer at the end of a run.
13. List the three most common anticoagulants used for FCI blood collection.
14. What is the maximum acceptable age of a specimen submitted for FCI analysis?
15. Outline a check list to be used when a specimen first arrives in the FCI laboratory.
16. Give the temperature range that is appropriate for transporting and storing specimens for FCI.
17. Name the two results obtained by hematology testing that are used by the flow cytometry laboratory.
18. Name the units with which flow cytometric results are expressed.
19. Name three methods for erythrocyte lysis.
20. What is the recommended solution for cell fixation in immunophenotyping?

VIII

Analysis of FCI Results and Troubleshooting

INTRODUCTION

After the immunophenotyping is performed, the results for each patient, each staining batch of samples, and each day's run must be evaluated by quality assurance standards. If QC specifications are not met, a troubleshooting routine should be followed. If specifications are met, however, clinical results can be reported.

NORMAL PROCESS CONTROL

For lymphocyte subset analysis, a normal whole blood specimen should be run as a process or method control each time FCI testing is performed. The normal blood specimen should be run after the flow cytometer is properly set up for immunophenotyping, the electronics have been standardized, and all the quality control checks have been passed. The procedure used to process the blood should be identical to that used for the test specimens. With proper selection of the normal specimen, the data obtained from the normal control can be used to develop and/or verify a reference range for the laboratory.

What is Normal?

One source of normal blood is a blood bank. By using a blood bank, normal specimens can be selected from persons who have no known medical problems. Specimens should not be obtained from individuals taking certain medications, pregnant women, or anyone with diseases or other conditions that might affect normal immune function.

Reference Range

Normal process control specimens should be selected to reflect the age, sex, and ethnic mix of the patient population being served by the laboratory. Many times, the reference range is composed of specimens from a wide variety of ages and ethnic backgrounds, without indication of specific demographic ranges. Any of the process control specimens used for the reference range should represent the same population as the patient population.

Obtaining specimens from the blood bank eliminates the possibility of confusing adult specimens with those for pediatric reference ranges. When obtaining the blood specimen, it is important to record all the factors that may affect the lymphocyte subsets including race, age, sex, exercise, infection, alcohol or tobacco use, and time of day that the blood was drawn.

To establish a reference range it is not sufficient to select and repeatedly test the same donors, for example, friends, relatives, or employees of the laboratory. Retesting the same donors may be used as a normal process control for quality control but may not be used to establish, to augment, or to verify the reference range.

SUPERVISORY REVIEW OF THE DATA

Flow cytometric data should be reviewed at two levels. The flow cytometer operator should provide the first level of review as data are being collected and while the FCI worksheet is being completed.

The second level of data review should be done by the flow cytometry supervisor, who should visually inspect the FCI data plots and statistics, and the hematology results, and should review all calculations before signing the report. The supervisor should also review a quality assurance checklist to ensure that the instrument was properly set up and operating on that particular day.

Immunophenotyping Worksheet

A flow cytometric immunophenotyping worksheet provides all of the appropriate information about the specimen. The worksheet should be completed as the data are collected or analyzed by the technician who performs the analysis. Important information on the immunophenotyping worksheet should include the percent lymphocytes in the specimen, percent lymphocytes included in the gate, proportion of cells nonspecifically stained using the isotype controls, the percent positive for each of the sample tubes, all relevant calculations, and the WBC and differential counts from the hematology laboratory.

The technician or technologist should determine if the patient immunophenotyping data meet quality control acceptance criteria. If the data are acceptable, the worksheet should be signed by the technologist and submitted for a final report. If the data are outside of the quality control specifications, either the specimen must be reprocessed or the sample tubes must be reanalyzed to determine the reason for failing the QC specification(s). The technologist should reanalyze the data following approved procedures.

The flow cytometry laboratory supervisor should review the immunophenotyping data and the immunophenotyping worksheet. The supervisor should visually inspect all the raw data, including the dot plots illustrating the appropriate gate set for light scatter, as well as the positions of the fluorescence markers for each of the subsequent tubes. The supervisor should review all of the numbers generated from the specimen and ascertain that the calculations are correct. The supervisor must also determine whether the immunophenotyping data meet quality control acceptance criteria.

Both the technologist and the laboratory supervisor need to review the immunophenotyping results for quality assurance. All results should be examined, including light scattering, marker settings, and population identification. An evaluation must be made of internal consistencies and logical relationships for the immunofluorescence panel.

It is important to detect sample outliers that are inconsistent with the other tubes in a specimen. The reason for the discrepancy should be identified and the sample should either be reanalyzed (using list mode data), rerun through the instrument, or rejected, in which case a new sample should be stained.

Problem samples must be identified and need to be distinguished from problem specimens. It is the laboratory supervisor's responsibility to determine whether the specimen must be rerun, restained and rerun, or rejected with a request for a new specimen to be submitted. Review of the immunophenotyping results always begins with an evaluation of the light scatter patterns to establish whether there was adequate discrimination between cell populations.

Quality Assurance Checklist

The flow cytometry laboratory supervisor must check to determine that proper quality assurance procedures were followed in the laboratory on that particular day. The supervisor should specifically check to see that the required standards were used for instrument setup and qualification and that the daily settings were comparable with those established previously for that particular instrument. A review of the Levy-Jennings plot should be made to see if the instrument has changed from its previous operating conditions.

The supervisor should ensure that a normal specimen was run at the beginning of the shift. The origin of the specimen should be reviewed to see if it is representative of the patient population. If it is representative, meets the quality control criteria, and has not previously been included, then it may be added to the database to expand the laboratory's reference range.

The supervisor must ensure that the hematology instrument was qualified according to quality assurance procedures. This requires that the flow cytometry laboratory and the hematology laboratory work together to obtain correct test results. Reproducibility criteria for the laboratory must be established. The supervisor must ensure that technologists running the FCI samples have received proper training. Variation introduced by the individual performing the test should be minimal and test results should be consistent from one technologist to another.

Light Scatter

Each specimen should be evaluated with respect to the ability to separate lymphocytes from other cells. Each specimen should also be assessed for the amount of debris and number of dead cells included. An inadequate specimen may require reprocessing with the entire reagent panel or a new specimen may need to be requested.

Comparing Light Scatter in Sample Tubes

The light scatter patterns for all samples from the same specimen should be compared (Fig. 8.1). All light scatter patterns should be identical for a single specimen. Differences in light scatter patterns from one tube to another suggest inappropriate processing or poor sample handling. If one of multiple samples is different in its light scatter pattern, the results for that sample are suspect: for example, Tube #5 in Figure 8.1.

Setting Lymphocyte Gates

For CD4+ T-cell quantitation, the position of the lymphocyte analysis gate should be reviewed (Fig 8.2). The total number of lymphocytes in the sample should first be determined. The total number of lymphocytes can then be compared with the proportion of lymphocytes included in the gate. The review process should specifically determine whether ≥90% of the lymphocytes were included in the light scatter gate. If the specimen is outside the QC specifications, the data should be reanalyzed using a new lymphocyte gate.

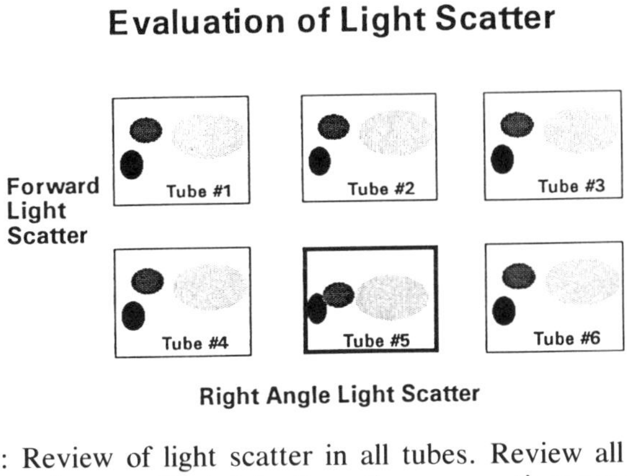

Fig. 8.1. QA: Review of light scatter in all tubes. Review all light scatter patterns for consistency. Each tube from the same patient specimen should have an identical light scatter pattern. Results from the outlier tube (#5) should be questioned.

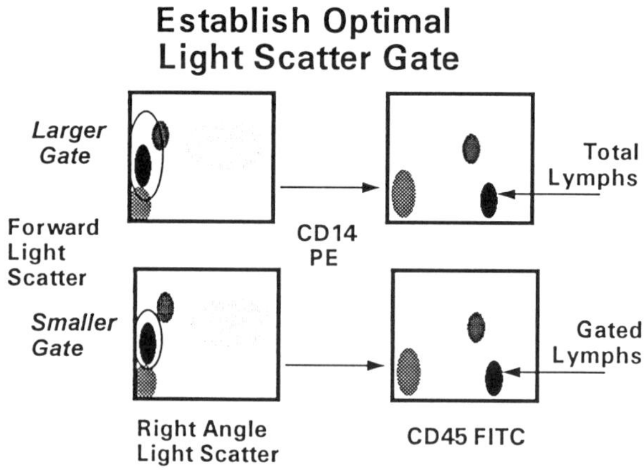

Fig. 8.2. QA: Review of light scatter gate. The gates are represented by ellipses. The light scatter gate for each patient specimen should be reviewed to ensure that the gate contained >90% recovery of lymphocytes and >85% lymphocyte purity.

The purity of the gate must also be determined in the review process. The gate used to analyze the subsequent samples must yield lymphocytes that are ≥85% pure in order to meet quality control specifications. It is also important to enumerate and to quantify the nonlymphoid cells that are included in the gate so that the results can be corrected. Knowing the number and type of nonlymphocytes in the gate is helpful when evaluating problem specimens.

Isotype Controls

The marker settings based on the isotype control should be checked for each specimen. The negative population should be easily identified as a tight, uniform population and fluorescence markers should be set so that positive staining is <2% of the cells.

The position of the negative population defined using the isotype controls should be identical to the position of the negative populations in all subsequent tubes. The position of the negative population should not shift between samples for a particular specimen. Most importantly, the negative population in any one sample tube from the same specimen should not move beyond the fluorescence marker established using the isotype control (Fig. 8.3).

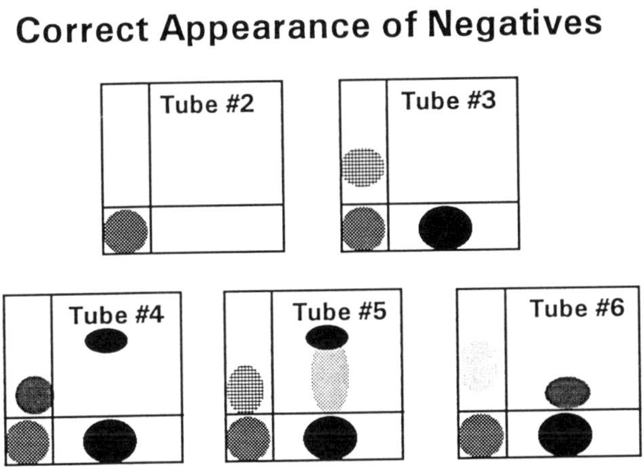

Fig. 8.3. QA: Consistency of negative population in the HIV immunophenotyping panel. The negative population in the isotype control tube (#2) should be in the same relative position as the negative population in each tube from that patient specimen.

Nonspecific Staining

Nonspecific binding is often observed as a diagonal population originating from the negative cells (Fig. 8.4). Markers should be set to ensure that these cells are recognized as negatives and not as positively stained cells.

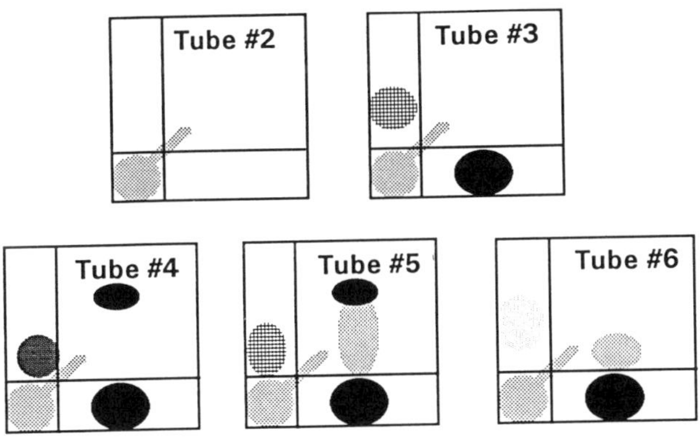

Fig. 8.4. QA: Nonspecific binding review in the HIV immunophenotyping panel. The nonspecific binding pattern is a diagonal smear from the negative population into the upper right quadrant. Markers may be changed to offset this nonspecificity as long as subset values are not affected.

Spectral Compensation

The proper compensation settings should be checked for each specimen (Fig. 8.5). As an example, the mean values of the CD3+ and the CD19+ populations must be at right angles to the mean value for the negative cells. The fluorescence markers established with the isotype control must not include any of the CD3+ or CD19+ cells in the double-labeled quadrant.

If the compensation is correct for the CD3,CD19 pair, the marker positions should also be checked for any remaining samples. The populations that are CD3+ in each of those tubes should be clearly identifiable. Any double-positive populations in the remaining tubes will also be CD3+. The cells expressing only CD3 should be identified using the marker set on the isotype control.

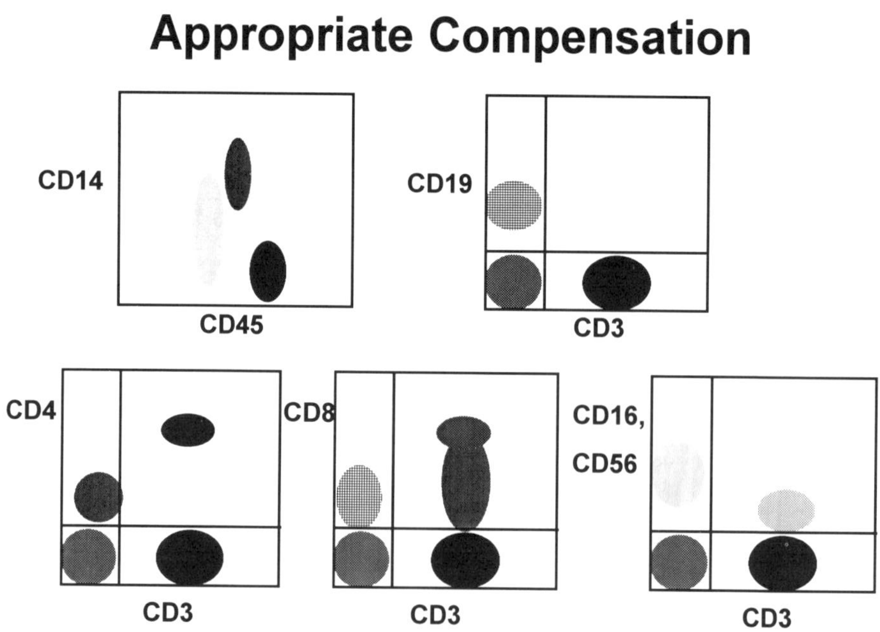

Fig. 8.5. QA: Compensation checks. Spectral compensation can be checked with the CD3,CD19 tube. If this is correctly compensated, all other tubes in the panel should be reviewed for consistency.

QUALITY CONTROL PARAMETERS

Quality control specifications should be designed to monitor whether the FCI results make sense in a logical way. Each patient should be evaluated for certain relationships, depending on the monoclonal combinations used in the test. For lymphocyte evaluations, a lymphocyte summation should be performed. For the HIV panel recommended by the CDC, replicates of CD3 should meet QC requirements.

> **Assumption: It is assumed that subset information from different samples of the same specimen make logical sense and that data from different reagent combinations can be logically combined.**

Logical Relationships

Other logical relationships, shown in Equations VIII.1, must also be met in order for the specimen to satisfy the QC criteria. For CD4+ T-lymphocyte quantitation, the subsets should have certain arithmetic relationships. The numbers of helper/inducer T cells and cytotoxic/suppressor T cells should equal the total number of T cells ±10%.

The CD3+CD4− population should be approximately equal to the CD3+CD8+ population and the CD3+CD4+ population should be approximately equal to the CD3+CD8− population. The CD3−CD19− population should be approximately equal to the CD3−(CD16+CD56)+ population.

$$(CD3+CD4+) + (CD3+CD8+) = CD3+$$
$$(CD3+CD4-) = (CD3+CD8+)$$
$$(CD3+CD4+) = (CD3+CD8-)$$
$$(CD3-CD19-) = CD3-(CD16+/CD56+)$$

Lymphocyte Summation

The summation of uncorrected lymphocyte percentages should equal the percent purity of the light scatter gate ±10%. The corrected lymphocyte percentages are obtained by dividing the percentage of each lymphocyte subset obtained from FCI by the purity of the light scatter gate and then multiplying by 100. The sum of the corrected percentages for T, B, and NK lymphocytes should equal 100% ±10% as shown below in Equations VIII.2.

% T Lymphocytes + % B Lymphocytes + % NK Lymphocytes = 100% ± 10%
Average % CD3 + % CD19 + % CD3−(CD16+ and/or CD56+) = 90–110%
(Corrected for Purity)

Replicate CD3 Determinations

The replicate determinations of CD3 should also be compared (Fig. 8.6). The intensity of staining for all CD3 populations should be identical. When CD3 FITC is used in multiple sample tubes from the same specimen, the intensity of CD3+ fluorescence staining should be the same for each of these tubes. The percentage of CD3+ cells determined for each sample in the panel must be within the laboratory range of reproducibility. The recommended range

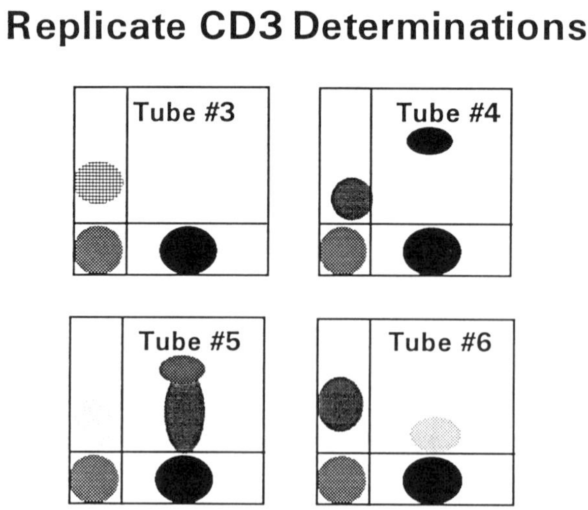

Fig. 8.6. QA: Replicate CD3 values. All replicate tubes for CD3 (or any other analyte) should be reviewed for comparable fluorescent intensity patterns. The CD3 values from replicate tubes should fall within a 5% range.

for the CD3+ results for each of the samples should not be greater than 5% between the highest and lowest value.

TROUBLESHOOTING

When samples fail to meet QC requirements, a troubleshooting procedure must be followed. It is important to determine whether the difficulty lies with the instrument, the reagents, or the patient. Resolution of the problem may require restaining, reanalysis, or resubmission of specimen.

Samples Failing Quality Control Specifications

Troubleshooting difficult specimens requires a careful analysis of how the sample fails QC specifications. Any sample that does not meet the QC specifications following the routine analysis procedures should be flagged. This assumes that a clear analysis procedure has been established and is followed in the laboratory. If a sample tube or specimen is flagged, the instrument settings and log book should be reviewed to see if the instrument has met the specifications for standardization.

Defined Reanalysis Procedures

A reanalysis procedure should be established to extract the maximum amount of information from any submitted specimen. The quality assurance program should define who collects the data and who reviews the results. If a sample does not meet specifications, it is important to determine who will perform the reanalysis.

List Mode Versus Real Time Data

For problem samples it is important to determine when the reanalysis will occur. If the specimen does not meet specifications, should the specimen be saved or thrown out? Are the data saved in list mode so that a supervisor or laboratory manager can then perform reanalysis, when needed, or are the data only analyzed in real time as the data is collected? Procedures should be established in each laboratory so that specimens not meeting quality control specifications can be appropriately reanalyzed by a qualified individual.

Specimen Versus Sample

The first stage in reanalysis of data focuses on assessing whether the problem is in a specific sample tube or within the original patient specimen received by the laboratory. If the specifications affect only a single tube, then the flag is indicated for the sample tube rather than for the specimen. If the replicate CD3 determinations are outside of the 5% range specification, the problem sample should be identified. If the remaining tubes meet all other QC specifications, then most of the data may be usable.

If the flagged sample is not the CD3,CD4 tube, then the CD4+ T-lymphocyte count can be reported. The light scatter patterns of the outlier sample should be compared with the light scatter patterns of the other tubes. Similarly, a comparison should be made between the outlier tube and the negative control in order to assess any reason for the sample deviation.

Poor Light Scatter Resolution

If the difficulty is because of a problem with the specimen, then all of the sample tubes will be affected. Poor light scatter resolution can also affect all of the data obtained. To separate populations, data should be displayed as forward light scatter (FS) by CD45 fluorescence and right angle scatter (SSC) by CD14 fluorescence (Fig. 8.7). Very large light scatter gates should be set to include the maximum number of lymphocytes rather than to maximize lymphocyte purity. Calculations may require significant correction factors for nonlymphocytes within the gate. Extra care is required in setting the light scatter markers in these instances.

It may be useful to display the data in this format whenever analyzing specimens with poor light scatter resolution. The combination of forward light scatter and CD45-FITC fluorescence provides an excellent means to define the lower boundary between debris and lymphocytes based on forward light scatter. This threshold must be established to maximize the lymphocytes included in the analysis window but minimize the contaminating debris (which is CD45−). Similarly, displaying CD14-PE fluorescence versus

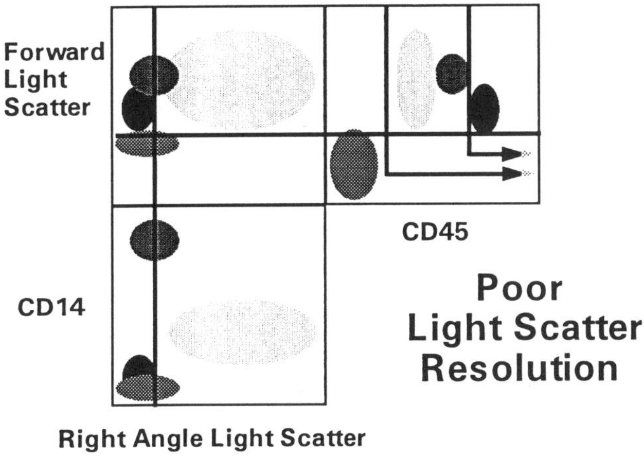

Fig. 8.7. Poor light scatter resolution. For specimens with poor light scatter, a very large light-scatter gate should be set. The forward scatter threshold should be drawn from the forward scatter versus CD45 histogram. The side scatter threshold should be drawn from the CD14 versus side scatter histogram. While lymphocyte recovery should be >90%, purity may be <85%.

right angle light scatter provides a means to separate lymphocytes from monocytes based on light scatter.

Nonspecific Staining

Certain specimens may have increased nonspecific staining. This is most readily observed in the isotype control tube (Fig. 8.8). Markers should be placed to exclude the majority of the double-negative cells from the positive populations. Markers should be set so that <2% of the events lie above the fluorescence boundaries in any of the positive quadrants.

Fluorescence Marker Movement Criteria

A major question in quality control of flow cytometric data is: "When can the fluorescent marker be moved to get correct results?" Clearly defined criteria must be established in every laboratory to define when and how fluorescent markers should be moved

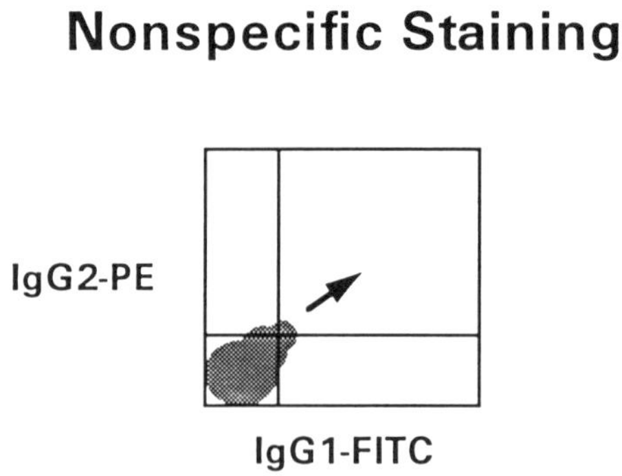

Fig. 8.8. Nonspecific staining. The fluorescence markers set on the isotype control should allow no more than 2% positive events in any positive quadrant.

from the position established using isotype controls. Figure 8.9 shows a case where fluorescence marker movement is indicated.

The criteria for moving the fluorescence markers should be based on whether the questionable events should be assigned to one population or another. For discrete populations, it is appropriate to move the boundary to eliminate the dim, nonspecifically stained cells. In this example, these events are clearly not part of the CD3+ population and should be included with the dim or negative cells.

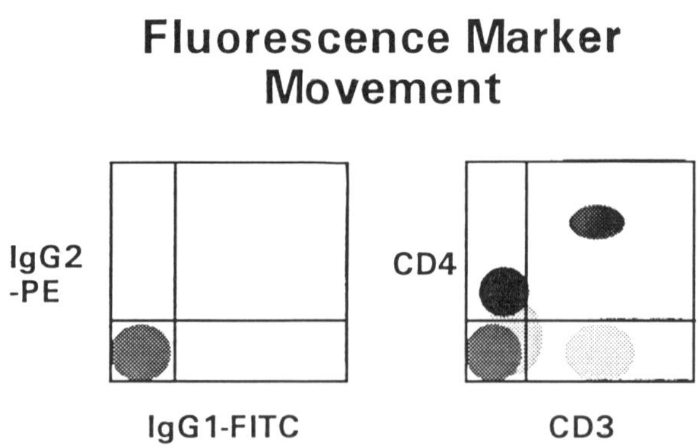

Fig. 8.9. Moving fluorescent markers. Fluorescent thresholds set on the isotype controls may be moved in cases where the positive populations are discrete patterns, easily resolved from the negatives.

For continuous populations the setting of the marker should be based on a reference population. The boundary setting may be more accurately established using the CD3,CD19 tube rather than the isotype control. This takes into account compensation settings as well as nonspecific binding to the negative cells.

Identifying the Problem

Identifying the reason for unacceptable sample or specimen quality is essential in the reporting of accurate flow cytometric results. Problems may occur from the flow cytometer hardware, reagents, inadequate specimens, random sample tube variability, or the flow operator's judgment in setting gates.

Hardware problems. Problems with hardware usually affect more than one specimen. The normal specimen used as a process control at the beginning of the run will also be affected by a hardware problem. A partial clog or bubbles in the fluidics can cause significant problems with a number of different specimens. Likewise, if there is an instrument sensitivity problem, there will be difficulty in separating the negative from positive populations for a number of specimens. In this case, the instrument standards should be rerun to ensure that the flow cytometer still meets the QC specifications.

Reagent problems. Reagent problems usually affect a single sample tube in the panel. The most common problem is the addition of reagents to the wrong tube or failure to add reagents to specific sample tubes. Such problems can be handled by running the recommended panel of sample tubes in the identical sequence every time. The patterns of correct staining become evident and the technologist can easily identify when there is a problem with reagents that have been improperly added. If one reagent combination exhibits dim staining for all the specimens being tested, then the normal specimen should be rechecked to see if the reagent performance is acceptable.

A titration of the antibody combination may show that the reagent is at an incorrect concentration. The new titration curve should be identical to the one obtained when the reagent was first purchased. If more than one reagent combination is affected, it is possible that the instrument or another common feature is the problem. It is useful to check the historical records of normals for the intensity of both the positive staining and nonspecific staining of the negative populations to see whether a reagent is still working properly. When in doubt, the reagents in question should be retested on a normal specimen.

Specimen problems. A problem with the specimen is usually detected by observing the light scatter patterns for the cell populations and the amount of nonspecific staining. A problem with instrument performance may be observed by checking the staining for the normal specimen. It is usually a good idea to keep the normal samples available during the entire run. The normal samples can then be rerun if difficulties arise, in order to show that the instrument is still operating properly.

Sample tube problems. Difficulties with individual samples can readily be identified. If one of multiple CD3+ determinations is out of range, then that particular sample is suspect. Often individual samples in a panel may have increased nonspecific binding or some abnormal reactivity pattern. It may be appropriate to change the marker setting for that tube if discrete populations are observed; however, the results should be flagged for reanalysis and review.

Gating problems. If there is a problem in gating, one or more samples may not meet specifications. If the proportion of lymphocytes is too low, important lymphocyte subsets may be lost. If too many nonlymphocytes are included in the gate, the results may also be biased. For these samples, the light scatter gate should be optimized and then the sample should be reanalyzed. If a CD4+ T-lymphocyte count cannot be determined, a new specimen should be requested.

Problem Specimens

Although problems may arise from a variety of sources, three types of problem specimens are often seen in FCI. Once the problem is identified, troubleshooting procedures may allow problem resolution if the results then meet QC standards.

Lymphopenic specimens. Certain specimens may present problems requiring more careful analysis by the laboratory. Specimens from lymphopenic patients must be handled with special care and consideration. Since fewer lymphocytes are present in these samples, more total cells must be analyzed. The increased proportion of nonlymphocytes that may be included in the light scatter gate may interfere with the results. It also may be more difficult to obtain an accurate WBC and automated differential on these specimens.

In specimens with low proportions of lymphocytes, more counts must be collected in order to maintain statistical precision. Lymphocytes usually constitute approximately 30% of leukocytes. A three-fold decrease in lymphocytes requires a three-fold increase in the total number of cells analyzed. For standard lymphocyte subset analysis, it is necessary to have counted at least 2500 lymphocytes for each sample. The increased total number of cells analyzed may be accomplished by increasing the total time the sample is run. It is also possible to concentrate the cells so that the final resuspension volume is decreased. For example, instead of resuspending the cells to 1 mL, a lymphopenic specimen can be resuspended to 0.5 mL.

A related problem can occur during bacterial infection. In response to infection, the total number of neutrophils may dramatically increase, relative to the total number of lymphocytes. Again, a total of 2500 lymphocytes should be counted for lymphocyte analysis. In this case also, the total number of cells must be increased in order to obtain 2500 lymphocytes for analysis.

Care must be taken to exclude or to correct for nonlymphocytes in the light scatter gate. The absolute amount of debris or nonlymphocytes may be the same but there are fewer lymphocytes. Therefore, the relative amount of gate contamination is increased, making analysis more variable. When a large correction for nonlymphocytes in the light scatter gate is required, it is important to check the light scatter patterns for each of the sample tubes in the panel to assess any changes from tube to tube.

For lymphopenic specimens, it is important to identify the basophils and correct for their presence. Basophils scatter light almost identically to lymphocytes. In normal specimens approximately 1% of the leukocytes are basophils and 30% are lymphocytes. Therefore, there are 30 lymphocytes for every basophil in a normal specimen. In lymphopenic specimens, however, the basophils may remain constant while there are relatively fewer lymphocytes. This means that in a lymphopenic specimen where the lymphocytes are decreased by a factor of three, the basophils comprise 10% of the cells in the light scatter gate. Basophils can be separated from lymphocytes on a CD45 by FSC scatter plot. Care must be taken in the analysis of lymphopenic specimens to identify and correct for nonlymphocytes in the light scatter gate.

Incomplete erythrocyte lysis. Occasionally there is incomplete lysis of the erythrocyte population. This may be a consequence of medication that the patient has taken, or it may be an inherent feature of the erythrocytes for that particular individual. If this occurs it is important to either change the lysis procedure or to correct the analysis for erythrocytes by careful electronic gating.

Certain samples may have a high proportion of debris or have erythrocytes that are resistant to lysis. For those samples in which the erythrocytes are not lysed, a second round of lysis may solve the problem. It is possible that the erythrocytes will be sensitive to a different lytic protocol that has also been validated by the laboratory and thus may be used as a backup.

With samples that contain a high proportion of debris, more counts should be collected since many of the events will be nonlymphocytes. It is important to determine the number of true lymphocytes based on CD45 expression and correct the data appropriately. This may result in a rather large correction of the percent of nonlymphocytes in the light scatter gate. In this case, it is important to compare the light scattering patterns for each sample tube in the specimen to ensure that they are identical.

Poor cell viability. Specimens with low cell viability also cause significant problems for FCI analysis. Nonlymphoid cells often change their light scattering characteristics when they die and may then be counted as lymphocytes. Nonviable cells often exhibit increased nonspecific staining that can interfere with the analysis. Specimens with significant loss of cell viability have preferentially lost one or more major subpopulations of cells from the analysis.

Large numbers of nonviable cells can increase the error in FCI results. Dead granulocytes and other cells may be included in a lymphocyte light scatter gate and therefore must be identified to correct the results. Nonviable cells may increase the double-negative population in certain reagent combinations or may cause difficulties in distinguishing dim from negative cell populations.

REPORTING IMMUNOPHENOTYPING RESULTS

Patient results may only be reported if supervisory review verifies compliance with quality control parameters. In the reporting process, the laboratory may also be called upon to assist the clinician with interpretation of the FCI results.

Patient Report Form

The patient report form must include all pertinent data from the result of the immunophenotyping test on the patient specimen. Patient information should include name, and/or patient ID, and other information that may affect interpretation of the immunophenotyping results. This information is used to determine the appropriate reference range for evaluating the results.

The panel of antibodies used in the immunophenotyping procedures should be listed by CD nomenclature. The relevant results and the laboratory reference range for each population of cells reported should also be included. All calculations of both percentages and absolute numbers should be shown on the report form. A telephone number must also be included for the contact person to answer questions or help in the interpretation of the data.

Antibody panel. The panel of antibodies should be complete enough for the clinician to understand and interpret the results. The antibodies should be listed using CD nomenclature. For paired reagents, the fluorescein conjugated antibody should be named first, followed by the phycoerythrin antibody. The commerical name or clone name should also be listed because subtle differences may be noted when different antibodies are substituted for each other. The cellular reactivity pattern normally observed with the particular antibody should be included in the report. This is to help educate clinicians and other professional staff who may not be familiar with the entire CD nomenclature.

All of the relevant populations should be easily identifiable. The percentage of each lymphocyte subset should be reported along with the absolute numbers calculated from the hematology results.

Other information important for demonstrating proper quality control should also be included: the percent of lymphocytes included in the gate, the purity of the gate including percentages and identity of nonlymphocytes in the gate, and the range of replicate CD3 determinations.

Reference range. The normal reference range for each lymphocyte subpopulation should be provided along with the patient results. This will help the clinician understand where the patient values fall relative to a normal reference population. The reference ranges for each lymphocyte subset must be established and/or verified for each FCI laboratory performing the CD4+ T-lymphocyte determinations. If published reference ranges are listed on report forms, the citation should also be printed.

Confidentiality issues. Patient confidentiality is an important consideration for all laboratory tests, but has become extremely critical for individuals who are HIV seropositive. The laboratory should never report patient results to unauthorized individuals and should have approved confidentiality procedures for the telephone reporting or facsimile reporting of patient results.

Physician Inquiries

Each laboratory must have established procedures for discussing the results with the ordering physician. It must be clearly defined who in the flow cytometry laboratory has the authority and responsibility to talk to the physician. Confidentiality is an important issue in HIV monitoring. This becomes crucially important in screening calls coming from a patient rather than the physician. A policy must be established for determining who is able to obtain results over the telephone.

The laboratory supervisor and/or director should establish a written laboratory policy for dealing with the following questions commonly asked by the physician:

1. Will you rerun the samples so my patient will qualify for treatment?
2. What confidence do I have in these results so that I can treat the patient?
3. Why do you need to run so many antibodies?
4. What is the importance of other subsets of lymphocytes in HIV infection?

5. What is the importance of the CD4/CD8 ratio?
6. How does CD34 quantitation compare with CFU assays?
7. What kinds of biologic variations can affect the results and the reference range?

Accuracy of cell enumeration. When a physician inquires about the accuracy of a flow cytometric result, what he or she is really asking is, "How good is this number?" The laboratory should have some data to support their precision of measurements at low, medium, and high values of each analyte. It may be more reasonable to report flow cytometric values as a confidence interval, although the misconception that the laboratory is not very accurate would undoubtedly result.

Influence of patient history and lifestyle on FCI results. In addition to the genetic variability of lymphocyte and lymphocyte subset levels, there are a number of lifestyle parameters that can also influence these analyses. Infections, pregnancy, smoking, exercise, and diurnal variation can all influence lymphocyte subsets and the laboratory should be prepared to discuss possible influences with requesting clinicians. Chapter VI presents a more complete discussion of these factors.

MAINTAINING DOCUMENTATION

Good laboratory practice requires documentation of all procedures performed on patient specimens. Documentation of reports as well as proof of laboratory proficiency should be available in the laboratory.

Procedure Manual

Procedure manuals must be written and maintained for all aspects of FCI procedures. These manuals must be modified or updated when any procedure is changed in the laboratory. Procedures for instrument setup must include primary and secondary standards as well as detailed procedures for alignment and electronic setup. Procedures for specimen collection, shipping, and storage must be included. Each step in the processing of a specimen, sample analysis, and data analysis must be clearly defined.

Procedures for review of the data and internal QC checks must be established. It is important to have flow charts of what to do when a sample does not meet quality control specifications. Procedures must be established for quality control of new reagent lots. Quality control procedures must also be established for the routine operation of the hematology instrument. This requires that the flow cytometry and hematology laboratories work together to establish these procedures. Procedures must be defined for establishing the reference range used by the laboratory. The laboratory reproducibility must be clearly defined and testing variability between technologists should also be assessed.

Log Books

As part of a quality assurance program, it is essential to keep record books of the important components of the FCI data. A log book must be kept showing instrument qualification and may also include Levy-Jennings plots. A record should be kept of patient ID numbers and relevant FCI data. This data will provide a reference of previous test results for individual patients. A log book should also be maintained for all the data used to establish the laboratory's reference range.

List Mode

List mode data should be collected on all samples run on the flow cytometer. This allows an invaluable saving of time and reagent cost when problems with sample analysis occur. A supervisor is able to regate and reanalyze problem samples to achieve accurate results without the need to reprocess specimens. How long the list mode data is retained is dependent on the laboratory's policies. Many flow cytometry laboratories retain list mode data until 1 week after patient results have been reported. Other facilities retain list mode data indefinitely, especially if therapeutic protocols are involved.

Hard Copy Results

At a minimum, all calculations used to determine the final results should be kept with the patient test results until the supervisor completes the record review. The calculations do not need to be included on the patient report form, however.

Any flagged samples or problems involved in obtaining the data must be noted and explained. If the data do not meet QC criteria, the reasons must be fully explained and their effect on the data noted. If any sample within the specimen is rejected, this must also be documented.

Retention of final report copies is mandated by regulatory agencies. For most state and federal requirements, hard copies (or comparable electronic copies) must be maintained by the laboratory for at least 2 years.

Proficiency Testing

A complete quality assurance program should be established at each laboratory. Each laboratory facility must be licensed by the appropriate state and/or local agencies. Laboratories should be inspected and approved by the College of American Pathologists (CAP). Inspections are required yearly for full accreditation. Some inspectors are very familiar with FCI, while others have little familiarity with the technology. Usually FCI is included in the overall laboratory inspection program.

Laboratories should participate in proficiency surveys when available. Proficiency testing has been mandated by CLIA-88. It is expected that the laboratory perform at an acceptable level established by the proficiency survey and that all records be available for review.

CONCLUSIONS

All flow cytometric immunophenotyping results must be reviewed for QA acceptability prior to reporting. Supervisory review must evaluate each histogram, all calculations, and logical relationships between samples. For CD4+ T-cell quantitation, lymphocyte recovery, gating purity, replicate CD3 values, and lymphocyte summation must also be reviewed and documented. If a sample fails a QC check, specific troubleshooting procedures must be followed to identify the problem and to determine whether the specimen should be restained, reanalyzed, or redrawn.

FURTHER READINGS

Bishop P, Boone D, Parker J. Immunophenotypes by flow cytometry: a longitudinal study in healthy individuals. *Diagnostic Clinical Immunology* 1988, 5:232–240.

Carter PH, Resto-Ruiz S, Washington GC, Ethridge S, Palini A, Vogt R, Waxdal M, Fleisher T, Noguchi PD, Marti GE. Flow cytometric analysis of whole blood lysis, three anticoagulants, and five cell preparations. *Cytometry* 1992, 13:68–74.

Centers for Disease Control. Guidelines for the performance of CD4+ T-cell determinations in persons with human immunodeficiency virus infection. *MMWR Morb Mortal Wkly Rep* 1992, 41(RR-8):1–17.

Giorgi J, Hultin L. Lymphocyte subset alterations and immunophenotyping by flow cytometry in HIV disease. *Clinical Immunology Newsletter* 1990, 10:55–61.

Jackson A. Basic phenotying of lymphocytes: selection and testing of reagents and interpretation of data. *Clinical Immunology Newsletter* 1990, 10:43–55.

Jackson A, Warner N. Preparation, staining, and analysis by flow cytometry of peripheral blood leukocytes. In Rose N, Friedman H, Fahey J, eds. *Manual of Clinical Laboratory Immunology,* 3rd ed. American Society for Microbiology, Washington, DC, 1986, pp 226–235.

Landay AL, Muirhead KA. Procedural guidelines for performing immunophenotyping by flow cytometry. *Clin Immunol Immunopathol* 1989, 52:48–60.

Loken M, Brosnan J, Bach B, Ault K. Establishing optimal lymphocyte gates for immunophenotyping by flow cytometry. *Cytometry* 1990, 11:453–459.

Malone JL, Simms TE, Gray GC, Wagner KF, Burge JR, Burke DS. Sources of variability in repeated T-helper lymphocyte counts from human immunodeficiency virus type 1-infected patients: total lymphocyte count fluctuations and diurnal cycle are important. *Journal of Acquired Immune Deficiency Syndrome* 1990, 3:144–151.

Margolick JB, Scott ER, Chadwick KR, Shapiro HM, Hetzel AD, Smith SJ, Vogt RF Jr. Comparison of lymphocyte immunophenotypes obtained simultaneously from two different data acquisition and analysis systems on the same flow cytometer. *Cytometry* 1992, 13:198–203.

National Committee for Clinical Laboratory Standards. *Clinical Applications of Flow Cytometry, Quality Assurance and Immunophenotyping of Peripheral Blood Lymphocytes* (H42-T). National Committe for Clinical Laboratory Standards, Villanova, PA, 1992.

Ohta Y, Fujiwara K, Nishi T, Oka H. Normal values of peripheral lymphocyte populations and T cell subsets at a fixed time of day: a flow cytometric analysis with monoclonal antibodies in 210 healthy adults. *Clin Exp Immunol* 1986, 64:146–149.

Reichert T, BeBruyer M, Deneys V, et al: Lymphocyte subset reference ranges in adult Caucasians. *Clin Immunol Immunopathol* 1991, 60:190–208.

STUDY QUESTIONS

1. Name two reasons for excluding a person from being a normal process control for FCI.
2. Describe how to determine the percentage of nonlymphocyte contaminants in the light scatter gate.
3. Describe how to use the CD3,CD19 sample tube as a quality control check for correct spectral compensation.
4. Describe how to use the negative isotype control tube to set fluorescence markers and give the maximum percentage of cells permitted above each of these markers.
5. Explain how to use CD3 staining in tubes 3, 4, 5, and 6 as an internal quality control check on the data in the CDC recommended HIV panel.
6. Give the number of specimens needed to establish a normal reference range.
7. Explain why it is better to set a light scatter gate that includes at least 95% of all lymphocytes and is only 88% pure than to have a gate of 98% lymphocyte purity that includes 91% of all the lymphocytes.
8. Describe how to do an additional quality control check for proper spectral compensation while running the sample tubes on the flow cytometer.
9. Describe the lymphocyte summation equation and how it can be used as a quality control check.
10. Describe the potential effect of each of the following on obtaining correct FCI results: lymphopenic sample, low antigen density, high contamination by nonlymphocytes, and excessive nonspecific staining.
11. Describe what to do when a specimen demonstrates considerable nonspecific staining and poor light scattering.
12. Describe a situation that would imply an instrument malfunction.
13. Name one of the most common problems in using monoclonal reagents.

14. Describe how the results from a normal process control are used to check the reagent staining in patient samples.
15. Describe what should be done if a specimen shows problems with light scatter and nonspecific staining while the normal control looks good.
16. Explain why it is important for FCI data to be reviewed by the flow operator prior to supervisory review.
17. Explain why is it important to have written procedures in the FCI laboratory.
18. Describe what to look for when evaluating the light scatter pattern for each sample tube in the panel.
19. List the information that should be included in the FCI report.
20. Describe the purpose of proficiency surveys.

IX

New Methods for Cell Enumeration

INTRODUCTION

There are many deficiencies and drawbacks in the current technology for cell enumeration by flow cytometry. The variability in results (both inter- and intralaboratory) is an outcome of the procedures used today for quantitative FCI. Multiple tubes are required to obtain the correct result. Multiple technologies are required to provide clinically useful results as absolute counts. The testing is expensive, requiring costly reagents and instruments. Operators of the instruments must be highly skilled, have specific training on each instrument, and must follow comprehensive quality assurance procedures to ensure the validity of the results. As a result, the testing is usually performed at high-volume laboratories or in a reference laboratory setting.

New techniques are being developed to extend or to replace the current flow cytometric methods. By directly determining the absolute counts of cell subsets on a flow cytometer, many of the difficulties described in this book are circumvented. The inherent problems of high cost and relatively slow throughput of the standard flow cytometric analysis has opened opportunities for methods for cell enumeration that are not based on flow cytometry. These techniques can directly or indirectly determine absolute numbers of cells with defined phenotypes per microliter, with equivalent or better precision than is currently obtained by flow cytometry. These new techniques are simplier to run, less expensive per test, and fit more readily into a routine clinical operation than the current flow cytometric analyses.

Although several innovations have been tested and used in research settings, none has been acclaimed as the new method of choice for enumerating cell populations. New procedures will be discussed in this chapter to illustrate the principles involved and the likely changes in routine laboratory workflow that must be anticipated as laboratories evolve from the current flow cytometric analysis to new procedures. This period of transition will require a thorough understanding of both current and new technologies and the relationship between clinical results from each.

INTEGRATION OF ABSOLUTE COUNTS WITH FLOW CYTOMETRY

As currently performed, flow cytometric results are reported in terms of percent lymphocytes stained. Historically, this relationship originated with the development of flow cytometric techniques in the immunology laboratory, rather than in hematology or laboratory medicine. Immunologists focused on the relative proportions of interacting lymphoid subsets, with little interest in the absolute quantitation of the cells.

The importance of absolute counts has become evident with the need to monitor peripheral blood cell populations in HIV disease or in PBSC harvests. The extension of immunologic data to the clinical laboratory requires the determination of the total number of cells in blood, which in turn necessitates the use of a cell quantitation procedure. This requirement has been solved by using data from two technologies, immunology and hematology, to calculate the concentration of cell subsets per unit of blood. The most direct method of eliminating the inherent problems with this approach is to modify the flow cytometers to provide absolute counts directly.

Techniques for Absolute Count Determinations

The determination of absolute counts requires enumeration of the cells of interest relative to a known volume of blood. This process can be accomplished by three different techniques: defined volume analysis, defined rate analysis, or analysis relative to a known number of added particles.

Both volume and rate analysis require carefully calibrated sample introduction fluidics on the flow cytometer (Fig. 9.1). During volume analysis, a precise amount of prepared specimen is removed from the tube and introduced into the flow cytometer. The sample introduction components are linked to the data acquisition computer so as to begin and end the collecting of data at the start and finish of a calibrated volume of injected sample.

Rate analysis is similar to volume analysis; however, the rate of injection of the sample (not the volume) is calibrated. Once the sample has reached a steady flow rate, the data acquisition computer is turned on to collect data for a precise period of time. The volume of sample interrogated by the laser can then be calculated to determine the number of events per unit volume.

These two techniques can be combined to provide internal quality control of the injection system. Calibrating both time and volume of the injection components provides two different ways of determining sample volume.

The third technique to determine absolute counts does *not* require the addition of precisely calibrated syringes for sample introduction. The addition of a known number of particles to a stained sample can provide a means of determining cell concentration (Fig. 9.2).

For this technique, the volume of sample, including the blood and added reagents, must be known and must remain constant from sample to sample. The added particles must be sufficiently different

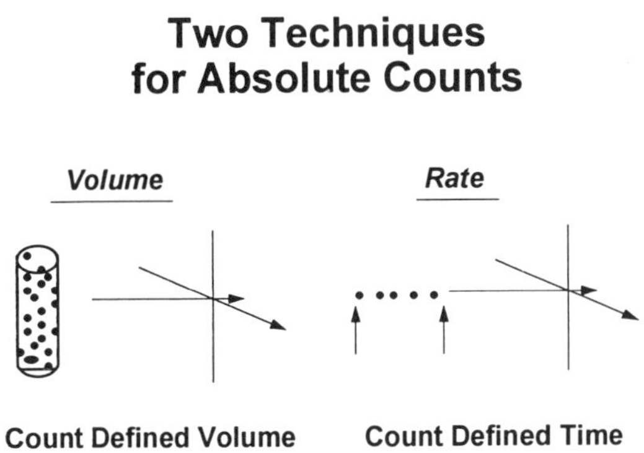

Fig. 9.1. Absolute counts by volume or rate. Absolute counts on a flow cytometer can be obtained by counting a known volume of sample, or by counting a sample flowing at a known rate for a specified time.

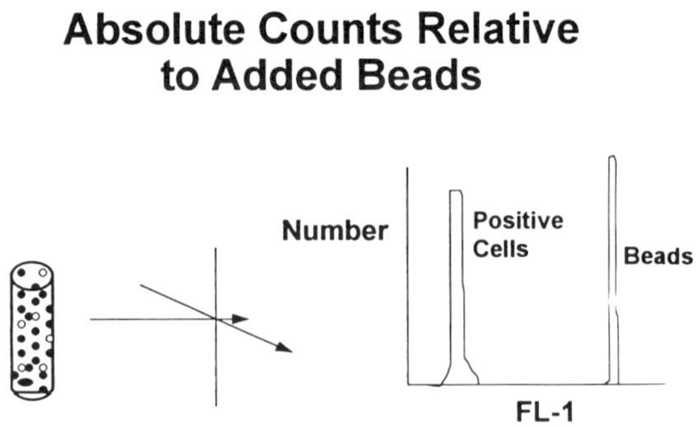

Fig. 9.2. Absolute counts by relative numbers. Absolute counts can also be determined by detecting positive cells relative to a known concentration of added fluorescent marker particles.

from the stained cell populations that they are not confused with cells. By determining the number of stained cells relative to the added particles, the concentration of cells can be calculated, based on the number of particles added to the total volume of sample.

Reagents Are Instrument Specific

All of these procedures require processing of the specimen by dilution only. The addition of reagents must be accomplished without centrifugation to maintain the relationship between the initial volume of blood and the concentration of labeled cells introduced into the instrument.

These significant restrictions placed on specimen processing are required to achieve appropriate labeling of the cells with sufficiently low nonspecific staining while lysing the erythrocytes and fixing the cells. The procedures are specific for the instrument for which they were developed. The protocols used for one instrument cannot be used for other flow cytometers without significant modification and validation. We do not recommend that the processing reagents from one manufacturer be used on instrumets made by other manufacturers.

Pipetting of Blood Defines Assay Precision

The precision and accuracy of these techniques depend on the measurement of a known volume of blood (usually 50 μL). All procedures require this step to be performed manually. Careful attention and training must be given to the reproducible dispensing of this small amount of blood. The specimen must be gently but thoroughly mixed. The micropipette must be calibrated with the appropriate set of tips. The tip must first be filled and the blood expelled to wet the pipet tip. Then, the blood must be slowly drawn into the tip and any residual blood on the outside removed, with careful attention to the biohazards involved. The blood should then be slowly dispensed into the very bottom of the staining tube, expelling all of the blood in the pipette. No blood can be left on the sides of the sample tube. New tips must be used for each specimen.

Additional Quality Control for Absolute Counts

The determination of absolute counts requires calibration of the sample introduction system or the known particles. Calibration of sample introduction depends on the use of calibration standards to assess the syringes and data acquisition timing. This procedure must follow the manufacturers' specifications and be performed at the start of every run. Keep in mind that the calibration particles must be thoroughly resuspended for proper use.

The calibration is usually performed after instrument setup, and optical and electronic standardization. At least two particle concentrations must be tested to determine that the system operates properly over the ranges expected for the test samples. A log of the results must be maintained and preferably recorded as a Levy-Jennings plot to visualize the variability over time. The best precision expected from the absolute counts is defined by the precision obtained from the replicate analysis of the calibration particles.

In the third technique, the concentration of added particles must be determined by an independent method, usually electronic cell counting. The added particles must be resuspended, diluted, and counted to determine their concentration. Reproducible pipetting of the particle suspension is critical to highly precise results. At least two concentrations of particles should be prepared for external calibration. A daily log must be maintained to record the variability of the counts and the reproducibility of resuspension. For those products where the particles are included with the processing reagents, it is essential to follow the manufacturer's recommendations for quality control of the reagents.

Advantages of Absolute Counts

The direct enumeration of cells identified by immunofluorescence circumvents many of the quality control problems currently encountered in flow cytometric analysis. The single-assay procedure eliminates the requirement for combining results from two different technologies. The calculation of absolute counts by multiplying the percentage of positive lymphocytes (determined by flow cytometry) by the number of WBC (obtained from hematology counters) is the greatest source of variability in determining CD4+ cell counts. The enumeration of the lymphocyte population by two different methods, FCI and hematology analyzers, which do not identify lymphocytes in the same way, is no longer a problem when absolute-count immunofluorescence is performed.

The entire approach to determining the number of cells identified by the immunofluorescence technique is changed when absolute numbers are obtained directly. There is no need to assess the proportion of cells stained relative to lymphocytes. The results are reported as number of positive cells per microliter, without a need to determine percent lymphocytes stained as an intermediate step. If the proportion of lymphocytes is an important parameter, this number can be calculated from the absolute numbers.

Equation IX.1:

$$\underbrace{\frac{\text{Positive Lymphocytes}}{\text{Total Lymphocytes}}}_{\text{Flow Cytometry; Light Scatter Gate}} \times \underbrace{\frac{\text{WBC}/\mu\text{L} \times \text{Lymphocytes}}{\text{Leukocyte}}}_{\text{Hematology; Differential}} = \underbrace{\text{Positive Lymphocytes}/\mu\text{L}}_{\text{Goal of Absolute Counts}}$$

This direct approach obviates the establishment of a lymphocyte light scatter gate for all lymphocyte determinations. If other cell populations, such as neutrophils, do not interfere with the identification of the fluorescence-positive cells, they do not need to be excluded by a light scatter gate. The difficulties in establishing optimal light scattering gates, maximizing lymphocyte recovery, and purity, become moot. The number of positive cells does not need to be determined relative to another cell population, but is determined per volume of initial test fluid. All gating and analytical procedures must focus on obtaining an accurate number of positive lymphocytes (numerator of Equation IX.1) without concern for the rest of the equation.

The direct detection of positive lymphocytes per microliter may be more stable over time. The current time limitations for storage of a specimen are related to the fragility of the neutrophils. These cells, not the lymphocytes, begin to die after 6 hours and limit the use of the hematology instrumentation to this time period.

Lymphocytes are longer-lived cells, which maintain their cellular characteristics for at least 48 hours at ambient temperature. If the destruction of the nonlymphoid cells does not interfere with the specific staining of the positive lymphocytes, the absolute lymphocyte count will not change over a period of days, even though the other cells may be lost. This quantitation will, however, be affected by clotting or other physical changes that decrease the number of the cells of interest.

Three-Color Analysis

The ability to obtain a direct absolute count permits the effective use of three-color immunofluorescence to reduce the total number of samples for a complete lymphocyte characterization. The same information obtained in 6 two-color immunofluorescence tubes is found in 2 three-color tubes. The subsets of all lymphocytes can be obtained with a combination of CD3 (T lymphocytes), CD19 (B lymphocytes), and CD56 (NK cells) (Fig. 9.3). These results are determined as an absolute number per microliter rather than as a percentage of lymphocytes. In Equation IX.2, the three values together constitute an accurate definition of the total lymphocytes in the blood specimen.

Equation IX.2:

$$\Sigma\ \mathrm{CD3}/\mu\mathrm{L} + \mathrm{CD19}/\mu\mathrm{L} + \mathrm{CD56}/\mu\mathrm{L} = \mathrm{Lymphocytes}/\mu\mathrm{L}$$

T-lymphocyte gating. In a similar manner, the subsets of T lymphocytes can be determined in a single tube. By combining CD3, CD4,

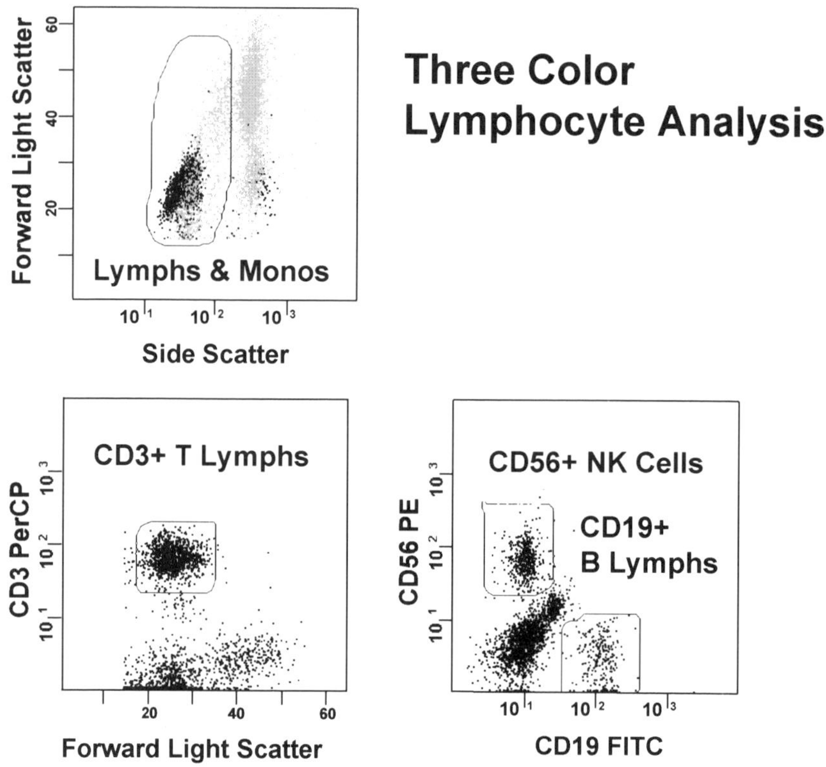

Fig. 9.3. Three-color analysis identifies T, B, and NK cells, each with a separate color. Neutrophils are excluded by using a large gate in forward and right angle light scatter. CD3+ T lymphocytes are identified in one color, CD19+ B lymphocytes are observed in a second color, and CD56+, CD3− NK cells are found in the third color.

and CD8 as three different colors, the absolute number and T-lymphocyte proportions of CD3+CD4+, and CD3+CD8+ cells can be found. The analysis is performed in a stepwise manner beginning with the identification of the CD3+ T lymphocytes (Fig. 9.4).

Gating on this discrete population of cells, when displayed versus forward light scatter, permits the further assessment of CD4 and CD8 as second- and third-color reagents. In this way the CD3+ lymphocytes can be discriminated from the smaller debris, and from the larger monocytes and neutrophils. The CD3+ gated cells

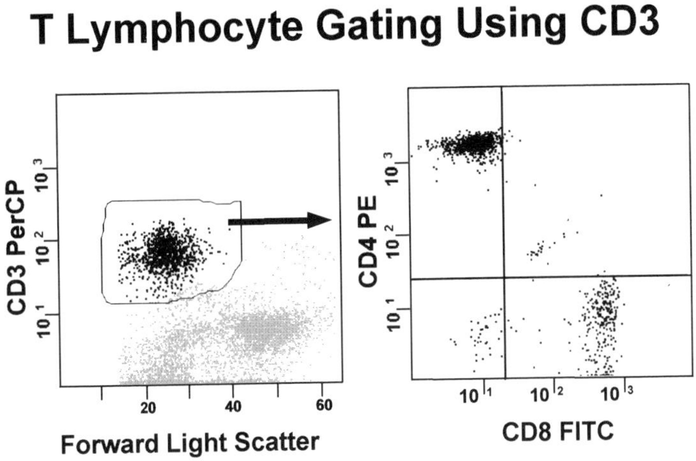

Fig. 9.4. Gating on T lymphocytes. T lymphocytes can be split into CD4+ and CD8+ populations by gating on CD3 in a three-color analysis. All cells in the right figure are CD3+ T lymphocytes.

are used to display CD4+ and CD8+. The CD4+ T lymphocytes are clearly a discrete population and distinct from the CD8+ cells, which include both dim and bright cells. The most accurate discrimination of CD8+ cells can be made by separating out the CD4+ cells. The minor cell populations of CD4−CD8−T lymphocytes and CD4+CD8+ T lymphocytes can be identified only by this three-color approach. The determination of absolute counts precludes the need for establishing a uniform and/or restrictive light scatter gate using CD45,CD14.

ALTERNATIVES TO FLOW CYTOMETRY

Several new techniques are being developed to quantify CD4+ cells without performing flow cytometric analysis. These new techniques simplify the lymphocyte subset detection and reduce the need for skilled technologists, while also reducing costs. None of these procedures has received wide enough use in clinical settings to allow a good comparison with the current techniques. We discuss these approaches to provide a more complete picture of the direction in which the technologies are moving in immunofluorescence enumeration of cells. Again, the transition from flow cytometric analysis to any new procedure will take a significant time period and require careful comparisons of the results from various techniques, with fully integrated quality control. An understanding of the new techniques will provide a basis upon which to make these comparisons. (While the principles underlying these new techniques are known, many of the details of the assays are proprietary and have not yet been disclosed publicly.)

Reacting Cells with Beads

The coupling of monoclonal antibodies to large beads instead of small dye molecules permits the detection of cells without using a flow cytometer. With this technique the conjugation of CD4 to a microscopically visible bead provides a means for visually counting those cells that react with the antibodies (Fig. 9.5).

The analysis is carried out by mixing a known volume of blood with antibody-coupled beads followed by lysis of the erythrocytes. Quantitation of the cells that bind the beads can be performed microscopically using a volumetric counting chamber (hemacytometer) and counting those cells which have beads bound to them. If the beads are sufficiently large, these cells can also be detected using an electronic cell counter. The cells with bound beads are identified as larger by the cell counter than either the cells without beads or the beads alone.

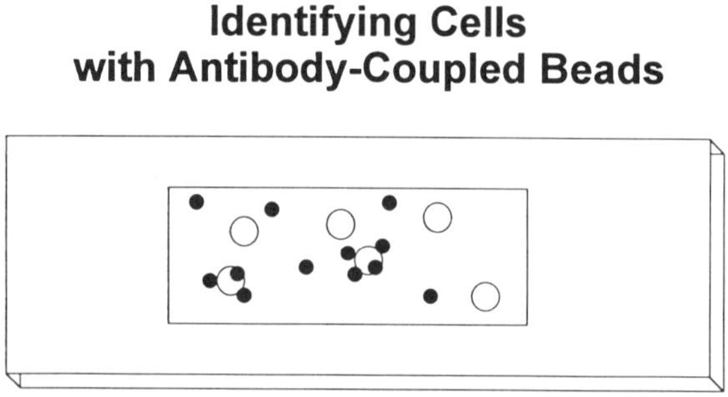

Fig. 9.5. Beads to identify cell populations. Antibody-coupled beads can be used to identify labeled cells visually in a volumetric counting chamber.

As with all absolute counting procedures, this technique requires reproducible pipetting of small volumes of blood and reagents. It is essential to discriminate the binding of CD4 beads to lymphocytes from the binding to monocytes. This can be accomplished by using a different-sized bead conjugated to CD14. Those cells having both large and small beads bound to their surface are considered monocytes and are ignored.

The microscopic detection of bead-cell conjugates must have a defined threshold number of beads bound per cell to be considered positive. This must be consistent and reproducible, even when different people score the results. For low counts, the number of microscopic fields required to identify the infrequent CD4+ lymphocyte may be too extensive, limiting the utility of the procedure. The use of an electronic cell counter has the same QC requirements as fluorescence analysis. These include instrument standardization, setting of the threshold to discriminate between positive and negative cells (as well as between lymphocytes and monocytes), and the calibration of the volumetric measurement.

Fluorescent Bead Detection

Fluorescent beads can also be used to quantify the CD4+ cells in whole blood preparations. The critical step in this procedure is discriminating between fluorescent beads bound to cells and unbound fluorescent beads. One approach has been to physically separate the CD4+ cells using magnetic microspheres. In this approach, a known volume of blood is reacted with a bead mixture containing CD4 antibody coupled to both fluorescent and magnetic beads. After a period of time to permit reaction, a magnet is used to attract the cell-bound magnetic beads (Fig. 9.6).

Fluorescent beads that are also bound to the CD4+ cells will be pulled to the magnet as well. The reaction tube can then be washed to remove erythrocytes and unbound fluorescent microspheres, leaving the magnetic microspheres and the cell-bound fluorescent beads. These remaining fluorescent beads can be quantified using a fluorescence ELISA (enzyme-linked immunosorbent assay) reader. This procedure is capable of processing large numbers of specimens simultaneously in a short period of time by a technician with limited technical expertise. It does not require the

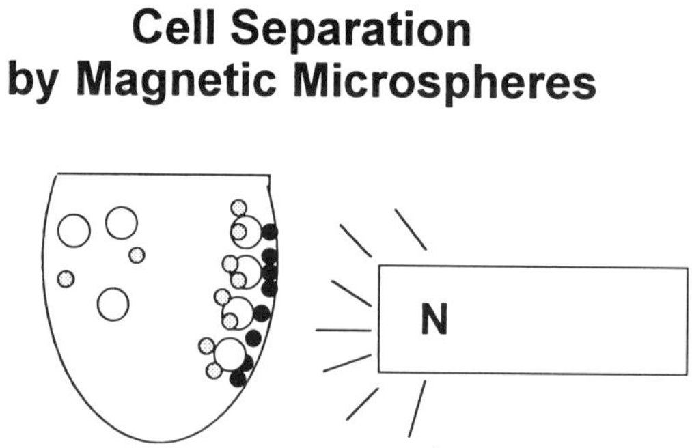

Fig. 9.6. Two types of beads to identify cell populations. Magnetic and fluorescent microspheres can be used together to quantify cell populations. Cells act as bridges to bind both magnetic and fluorescent beads. The magnetic beads are isolated by a magnet as the unreacted fluorescent microspheres are decanted with the unlabeled cells. Total fluorescence from each well correlates to total cells in the original sample.

sequential analysis of individual cells but provides a fluorescence number related to the number of CD4 cells in the initial sample.

The quality control issues in this assay are similar to those for other fluorescence assays. The calibration of the ELISA plate reader is extremely important for validation of the technique. The relationship between fluorescence signal and cell number must be checked at the beginning of each run. The discrimination between CD4 lymphocytes and monocytes must be integrated into the design of the bead system, since both the fluorescent and magnetic beads are conjugated with anti-CD4.

CD4 Antigen Quantitation

The amount of CD4 per lymphocyte is relatively constant so that by solublizing the cell membranes, the quantitation of CD4 antigen can be performed by ELISA. The amount of CD4 antigen can then be related to the number of cells that release that antigen. Several assumptions are made in this assay procedure. CD4 must be solublized in exactly the same quantitative way for every patient and control. CD4 from monocytes must not interfere or affect the standard curve for CD4+ lymphocytes. Importantly, soluble CD4 must not be present in the serum as this additional source would cause an overestimation of the number of CD4-bearing cells.

Another approach to antigen quantitation is a microtiter plate immobilization of antigen-bearing cells. Using monoclonal antibodies, cells can be simultaneously immobilized by "immunocapture" using one antibody and identified by a second monoclonal antibody that is enzyme-labeled. Results read on an ELISA plate reader are expressed as picomoles per liter of soluble antigen and can be related to cells per liter of blood.

This approach does not count cells, but determines a surrogate analyte, CD4 antigen, which is related to CD4+ cells. The possible errors introduced by differences between patients must be considered in validating the relationship between antigen amount and cell number. There are no reports of large studies to demonstrate that this relationship is valid for the majority of patients with a variety of secondary diseases and treatments.

These assays must be standardized in the same ways that ELISA or immunoassays are controlled. Calibration curves as well as positive and negative controls must meet quality control specifications for each run. This methodologic approach does not require highly skilled technologists and is adaptable to automation and batch processing of large numbers of specimens.

Two-Color Image Analysis

Cells labeled by immunofluorescence can be detected by automated microscopy as well as by flow cytometry. Absolute counts can be made directly by analysis of the images of cells contained within a precise volumetric cuvette (Fig. 9.7). The discrimination between positive and negative cells can be facilitated by using red laser excitation, where erythrocytes and leukocytes are invisible unless they bind a fluorescent dye.

Scanning a cuvette by moving both the laser and microscope stage can be accomplished within the same time period as analyzing the same number of cells by flow cytometry. Two-color immunofluorescence can be used to identify the CD3+ T lymphocytes and the simultaneous expression of CD4 or CD8 on these cells. By passing the laser beam several times over the same cell, it is possible to increase the signals of the positive cells relative to the back-

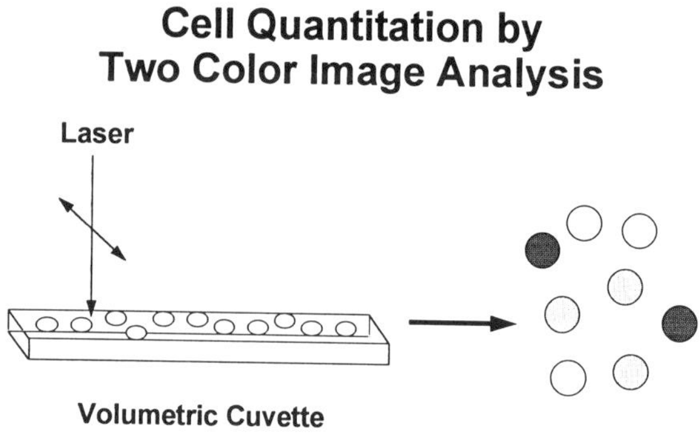

Fig. 9.7. Two-color immunofluorescence imaging. Cells labeled by two colors of immunofluorescence are introduced into a volumetric cuvette by capillary action. Positive cells are identified by scanning the cuvette with a HeNe laser beam. Red blood cells are invisible to this system.

ground. Appropriate dilution of the sample assures minimal spatial overlap between positive cells, and two-channel detection of positive cells reduces the false-positive detection rate.

The use of a cuvette to determine volume replaces the pipetting of blood in the analysis and reduces the measurement error. By integrating the reagents into the cuvette, pipetting of antibodies can also be eliminated. This approach both reduces potential pipetting errors and allows less skilled personnel to perform the test. Quality control of the entire procedure must consider not only the test itself, but also the instrument setup, since the operator has no ability to change or modify the settings. Standards and process controls are integral to the analysis and must be performed on a routine basis.

CONCLUSIONS

Immunophenotyping by flow cytometry is a complex, expensive technique for enumerating cells. In order to simplify the process and to reduce analytical variability from hematology measurements, all of the flow cytometer manufacturers are currently developing systems for the direct measurement of lymphocyte subsets.

Alternative technologies, utilizing ELISA, fluorimetry, and image analysis, have been proposed to replace flow cytometry in the routine clinical laboratory. Although the new techniques promise higher throughput, lower cost, and the requirement for less highly trained staff, both the procedural validation and the clinical correlation of these techniques are still under development. Regardless of which new technologies eventually replace flow cytometric immunophenotyping as a means of lymphocyte subset quantitation, their QA and QC procedures must address the same issues of instrumentation, reagents, and process controls.

FURTHER READINGS

Carriere D, Fontaine C, Bertheir A-M, Rouquette A-M, Carayon P, Laprode M, Juillard R, Jansen A, Paoli P, Paolucci F, Gros P, Pau B. Two-site enzyme immunoassay of CD4 and CD8 molecules on the surface of T lymphocytes from healthy subjects and HIV-1-infected patients. *Clin Chem* 1994, 40:30–37.

Check WA. Simplifying CD4 lymphocyte testing. *CAP Today* 1993, 16–25.

Denny TN, Jensen B, Garcia A, Vella F, Wong W. Determination of immune subsets by a new alternative technology. Abst #598, First National Conference on Human Retroviruses and Related Infections, Dec 12–16, 1993, Washington, DC.

Gaumer RH, Fink D, Jensen B, Vella F, Wong W. T cell subset quantitation by non-flow cytometric method. Abst #688, First National Conference on Human Retroviruses and Related Infections, Dec 12–16, 1993, Washington, DC.

Hughes MD, Stein DS, Gundacker HM, Valentine FT, Phair JP, Volberding PA, Within-subject variation in CD4 lymphocyte count in asymptomatic human immunodeficiency virus infection: implications for patient monitoring. *J Infect Dis* 1994, 169:28–36.

Landay A, Ho JL, Hom D, Russel T, Swerner R, Minuty JG, Kataaha P, Mmiro F, Jackson B. A rapid manual method for CD4+ T-cell quantitation for use in developing countries. *AIDS* 1993, 7:1565–1568.

STUDY QUESTIONS

1. Describe why new methods for cell enumeration are desirable.
2. Why is it desirable to eliminate the hematology measurement for absolute counts?

3. List three techniques that could be used to obtain absolute counts directly from the flow cytometer.
4. How are these techniques similar?
5. How are these techniques different?
6. What quality control is required for each technique?
7. Explain how all lymphocyte subsets can be measured by three-color analysis.
8. Describe how T-gating is performed.
9. What are the advantages of assays that are not based on flow cytometry?
10. What are the disadvantages of counting bead-bound cells by microscopy?
11. Describe how a magnetic field can be used to separate cells bound to magnetic microspheres.
12. What QC issues are important for the fluorescence plate assays?
13. Describe two approaches to measuring CD4 by microtiter ELISA.
14. What are the disadvantages of measuring soluble CD4 antigen?
15. Describe the use of a volumetric cuvette for two-color image analysis.

Index